# Archaeology of Britain s Oldest Church Doors

# Archaeology of Britain's Oldest Church Doors

Westminster, Hadstock and 'Dane-skins'

*by*
Warwick Rodwell

*With contributions by*
Krista Blessley, Ruairidh Macleod,
Peter Massey *and* Paul Reed

OXBOW | books
Oxford & Philadelphia

Published in the United Kingdom in 2025 by
OXBOW BOOKS
81 St Clements, Oxford OX4 1AW

and in the United States by
OXBOW BOOKS
1950 Lawrence Road, Havertown, PA 19083

Paperback Edition: ISBN 979-8-88857-229-0
Digital Edition: ISBN 979-8-88857-230-6

A CIP record for this book is available from the British Library

Library of Congress Control Number: 2025942524

Printed in the United Kingdom by Halstan & Co Ltd.

For a complete list of Oxbow titles, please contact:

UNITED KINGDOM
Oxbow Books
Telephone (0)1226 734350
Email: oxbow@oxbowbooks.com
www.oxbowbooks.com

UNITED STATES OF AMERICA
Oxbow Books
Telephone (610) 853-9131, Fax (610) 853-9146
Email: queries@casemateacademic.com
www.casemateacademic.com/oxbow

Oxbow Books is part of the Casemate Group

A generous grant towards the cost of producing this volume was made by the Dean and Chapter of Westminster

*Front cover: Westminster Abbey. The oldest church door in Britain, reused in the chapter house vestibule. Malcolm Crowthers, © Dean and Chapter of Westminster*
*Frontispiece: Hadstock church, Essex. North nave doorway. © Paul Ravenscroft*
*Back cover: Westminster Abbey. Reconstruction of the original appearance of the oldest church door. Warwick Rodwell, © Dean and Chapter of Westminster*

The Publisher's authorised representative in the EU for product safety is Authorised Rep Compliance Ltd., Ground Floor, 71 Lower Baggot Street, Dublin D02 P593, Ireland.
www.arccompliance.com

# Contents

# Contents

# Preface

My interest in the archaeology of church doors was aroused in 1973, when I was invited to direct the excavation of the interior of Hadstock church, on the Essex–Cambridgeshire border, preparatory to a new floor being laid throughout the nave and transepts. The north nave door was notorious for its presumed great antiquity and the fact that its exterior had once been covered with hide. The latter had all gone, but the legend persisted that it was the skin of a Dane who had been caught in the act of sacrilege, and flayed. The excavation took place in 1974, but archaeological recording of the fabric and research into the architectural history of the church continued intermittently for a further decade. The nave door and the decorated Romanesque arch in which it hangs were seminal components of that research, and the legend of the 'Dane-skin' needed objective clarification.

The late Cecil Hewett, a carpentry historian, was searching the country for evidence of the survival of Anglo-Saxon church woodwork, and set his sights on the Hadstock door as the most likely candidate. We studied this together and I invited a colleague from Oxford, the late Dr John Fletcher, to examine the possibility of scientifically dating the door by dendrochronology.

Hewett and Fletcher also became keenly interested in another door, possibly older than that at Hadstock. It hangs in the chapter house vestibule at Westminster Abbey, where it gives access to a medieval cellar. In 1999, I was commissioned by English Heritage to write a new guidebook to the chapter house and the adjoining Pyx Chamber, causing me to examine the battered and seemingly insignificant door which thousands of visitors walk past every day, mostly not even noticing it. Moreover, this door had formerly been covered with skin on both faces, which had been claimed as human, and variously attributed either to a Danish raider, or to the English merchant who instigated an audacious robbery in 1303 from the royal treasury, housed in the abbey cloister.

Remains of hides on church doors have also been reported by antiquaries at several other locations in England, since the mid-17th century, most importantly at Hadstock and Copford (Essex). Again, they had been claimed as human, and attributed to pillaging Danes. There were several cogent arguments why this gruesome claim could not be true: first, all the skins were applied to the doors during their construction in a carpenter's workshop, and not added subsequently; secondly, the doors to which they were attached post-dated the era of Danish incursions; thirdly, it was apparent that at least some of the skins were animal hides; fourthly, flaying was never a prescribed form of punishment in Church law or English civil law. Clearly, scientific investigations were needed to separate fact from fiction. Moreover, what was the purpose of covering doors with hide? Many questions needed to be addressed, starting with accurately dating the doors.

At the end of the last millennium, an important development in the application of dendrochronology (tree-ring dating) by Dr Daniel Miles made it possible to date the oak boards from which church doors were constructed. Architectural and archaeological evidence pointed to three doors as potential claimants for the status of being the oldest in Britain, and dendrochronology duly ranked them in date-order: Westminster Abbey, 1050s; Hadstock church, 1060s–70s; and Rochester Cathedral, *c.* 1080s–90s. The last did not have a covering of skin.

At the University of Cambridge, a seminal development in DNA studies made it possible to identify the animal species used for historic parchment and vellum documents and, by extension, for the hides on church doors. Few samples of these hides have survived to the present day, and their collagen was investigated at Cambridge in 2020 by Dr Ruairidh Macleod: Westminster and Hadstock are both confirmed as cow hide, and the skin from Copford church is horse/donkey. The results from the Worcester hide were obtained in 2025 and also proved to be horse/donkey.

Having been responsible for archaeological investigations that involved the doors and their settings at Westminster and Hadstock, it seemed logical to bring them together, along with such evidence as could be gleaned about other early church doors, regardless of whether they are known to have been covered with hide. Consequently, the major part of this book concerns the form, construction and decoration of the two earliest surviving English doors. A surprising variety of techniques is displayed in the later 11th and 12th centuries, and the Westminster door is unique. Its form of construction is unmatched by any other recorded door in Britain, prompting the question: are its origins Anglo-Saxon or Norman-French? Serendipitously, it also happened that two conservation carpenters and woodwork historians, Peter Massey and Paul Reed, requested permission to undertake a detailed study of the construction method and tools required to fabricate the Westminster door, and a chapter has been devoted here to their findings.

In sum, this volume reviews the construction and decoration of our oldest surviving church doors, and discusses also the tools and carpentry techniques involved. An uncertain, but seemingly small, proportion of 11th- and 12th-century doors were given a covering of animal hide on the exterior, which was then painted bright red before fixing the ornamental ironwork. On some doors, including an early one at Rochester Cathedral, paint was applied directly to the timber, without an intervening layer of hide. After examining the physical evidence and folklore relating to hide-covered doors in England, I have concluded how their false association with the Danish incursions of the 9th, 10th and early 11th centuries is likely to have come about.

Notwithstanding the pioneering research carried out initially by Cecil Hewett and latterly by Professor Jane Geddes, many questions about the history of church doors remain to be answered, and the field of investigation needs to be widened, with systematic searches conducted to identify more potential examples of doors of the 11th to 13th centuries, not only in Britain but also in Scandinavia and France.

Warwick Rodwell
Westminster Abbey
Easter 2025

# Acknowledgements

The material presented here is the result of fifty years of intermittent study of the church doors at Hadstock and twenty-five years at Westminster Abbey, during which time the writer has become indebted to many colleagues and acquaintances for their kind help and support.

The Dean and Chapter of Westminster, and their staff, encouraged and supported the study of the unique 11th-century door that hangs in the chapter house vestibule: in particular, the Very Rev'd Dr Wesley Carr, KCVO† and the Very Rev'd Dr John Hall, KCVO (successive Deans), John Burton, MBE and Ptolemy Dean, OBE (successive Surveyors of the Fabric), Major General David Burden, CVO, Sir Stephen Lamport, GCVO and Paul Baumann, CBE, LVO (successive Receivers General), and Jim Vincent† and Ian Bartlett MVO (successive Clerks of the Works).

For their unstinting assistance with research, I must thank Dr Tony Trowles (Head of the Abbey Collection and Librarian), Dr Richard Mortimer and Dr Matthew Payne (successive Keepers of the Muniments), Miss Christine Reynolds (Assistant Keeper of the Muniments), and Vanessa Simeoni (Head Conservator) and her colleagues. The staff of English Heritage were most supportive too, and funded the dendrochronology programme; thanks are especially due to Dr Steven Brindle, Dr Jeremy Ashbee and Derek Hamilton for their support. Dr Daniel Miles and Dr Martin Bridge carried out the investigation, and the initial elevation drawings of the door were drafted by Angela Thomas.

This study of the Westminster door has been considerably enhanced by the research and experimentation carried out by woodwork historians Peter Massey and Paul Reed, and I am grateful to them for agreeing to the publication of their investigations here in chapter 4. They wish to acknowledge the following: Westminster Abbey's Conservation team for providing lighting and equipment, so that they could carry out their study effectively; Dr Daniel Miles and Dr Martin Bridge for discussions about the door; Dr Damian Goodburn, in correspondence, regarding his theory on early medieval woodworking, and Dr Christine Rauer, Reader in Medieval English, University of St Andrews. Katie Meheux, University College London, Department of Archaeology Library, kindly sourced references during the Covid-19 pandemic. Thanks are also due to Hull and East Riding Museum for permission to use the image of the Viking chisel, and to Robert Williams who kindly shared his photograph of the Ebbsfleet plane.

Study of the door at Hadstock was facilitated by the rector and churchwardens; the Very Rev'd Michael Yorke† (Rector 1974–78) was immensely supportive, as has been Miss Patricia Croxton-Smith, the indefatigable village historian, throughout the half-century of my involvement there.

Of the many colleagues who have contributed to the study of Hadstock special mention must be made of Professor Eric Fernie, CBE, Dr John Fletcher†, Professor Jane Geddes, Dr Richard Gem, OBE, Adrian Gibson†, Cecil Hewett†, Kirsty Rodwell and Dr Harold Taylor, CBE†. A major debt of gratitude is due to Dr Ruairidh Macleod, then at the University of Cambridge, for his research into identifying the animal species represented by the so-called 'Dane-skins'.

My appreciation of the Rochester Cathedral door was enhanced through discussion with Tim Tatton-Brown, OBE and research carried out by Jane Geddes. Indeed, my debt to the latter from 1974 to the present day is incalculable, and her magisterial volume on medieval decorative ironwork has been the touchstone of my research on church doors for decades.

For arranging access to the detached door and assisting on site at Elmstead church I am grateful to Andrew Downton (Churchwarden) and parishioners Anthony and Jane Bedford. At Copford church, Claire Macaulay (Treasurer) kindly arranged access to the skin samples and rendered practical assistance with the study of the Norman door. Glynn Davis (Senior Collections Curator) of Colchester Museums helpfully made available the Copford sample of skin in the museum collection.

For access to the artefacts from Hadstock I am indebted to Jenny Oxley, Acting Curator of Saffron Walden Museum. Thanks are due to the Dean and Chapter of Worcester, and Dr David Morrison (Cathedral Librarian and Archivist) for permission to study the fragments of hide from the north doors, and to Fiona Keith-Lucas (Cathedral Archaeologist) for making the practical arrangements, supplying photographs and invaluable assistance with research. Dr Katie Miller (Archaeology Collection Curator, Hereford Museum) made the fragment of hide from Pembridge church available for study and supplied a photograph. Bruce Simpson at the Royal College of Surgeons very kindly arranged consent to reproduce the College's portrait of Professor J.T. Quekett; and Dr John Crook and Paul Ravenscroft generously gave permission to publish their excellent photographs of the Rochester, Hadstock and Copford doors, respectively.

I wish to record my deep gratitude to my wife, Diane Gibbs, for her customary unfailing support whilst making field trips to churches, and during the months that it has taken to prepare this volume. Finally, I gratefully acknowledge Oxbow Books for taking on the publication, and in particular Dr Julie Gardiner, Jessica Hawxwell, Mette Bundgaard and their editorial and production teams.

# 1

# Doors and 'Dane-skins'

Church doors rank amongst the oldest and most numerous timber artefacts surviving from the Middle Ages, but the archaeology of medieval doors in the British Isles has received less attention than it deserves. Those that have survived the ravages of time are almost entirely found in England: the sole example of note in Wales is at Gileston (Glam.),[1] none appear to have been recorded in Scotland, and only three in the Channel Islands. The locations of doors discussed in this study are shown in Figure 1.

## Historiography

While the presence of historic doors – especially those that bore decorative ironwork, or *ferramenta* – is often noted in antiquarian writings, and occasionally accompanied by line drawings, typological and regional studies are lacking. Some antiquarian treatises devoted a brief section to doors,[2] but it was only in the mid-20th century that they began to be seriously recorded and published. The instigator of a new, in-depth approach to the study of church carpentry was Cecil Hewett, who was especially interested in recording and dating the major structural components.[3] However, his three seminal publications also embraced many parish church and cathedral doors: *Church Carpentry* (1974), *English Historic Carpentry* (1980) and *English Cathedral and Monastic Carpentry* (1985).[4] Hewett was the first person to evince a serious interest in the technology of the jointing displayed by medieval woodwork, revolutionizing the dating of English timber-framed buildings.

One of the most elaborately decorated medieval doors is at Stillingfleet (N. Yorks.) and has been mentioned in numerous publications from the mid-19th century onwards, but it was only in the early 1970s that a thorough study of this door and its ironwork was tackled, resulting in an exemplary publication (Fig. 13).[5] This set a new

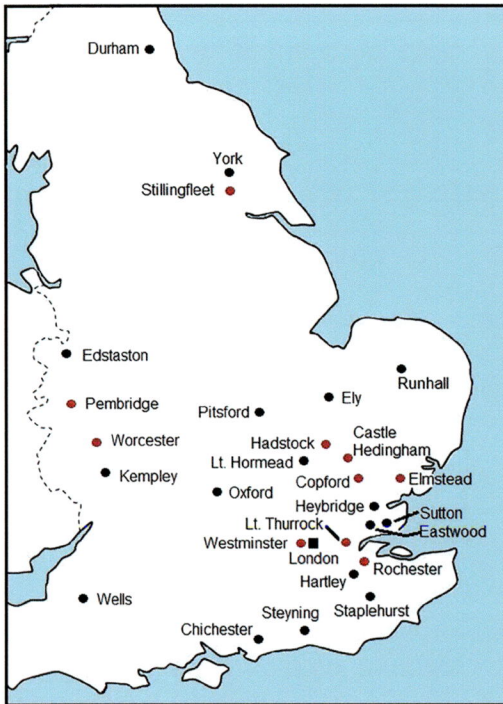

*Figure 1: Map showing the locations of the principal churches discussed in the text. Those marked in red formerly had hides attached to one or more doors. Author*

standard for the recording of church doors and their decoration. Meanwhile, in 1973, Jane Geddes began to research the subject of medieval decorative ironwork, and in 1999 published her magisterial corpus: *Medieval Decorative Ironwork in England*.[6] A high proportion of the ferramenta described relates to church doors, the constructional details of which have also been recorded by Professor Geddes.

Over the past half-century, the development of the science of tree-ring dating – dendrochronology – has revolutionized the accurate dating of oak structural timbers, panelling, doors and furniture. Dendrochronology has facilitated close, and sometimes very precise, dating of church doors, enabling us to place the few surviving later 11th-century English doors in chronological order.[7] Only one is assignable to the 1050s, thereby identifying it as Anglo-Saxon, and that is in Westminster Abbey (Fig. 10).

Although the Westminster door is the focus of this volume, it does not exist *in vacuo*, and study of its historical, archaeological and technological context is no less important. Additionally, this door and several others in diverse parts of England are notable for having been covered on the outer face with skin or hide, a few fragments of which still remained in place into the 19th century. Sometime before the 17th century, it was popularly surmised that these skins were human and belonged to Viking pirates who had been caught desecrating churches: they were flayed and their hides nailed to the doors as a warning to other potential perpetrators of sacrilege. Thus the legend of the 'Dane-skin' was born and nurtured in those parishes where there was evidence for a church door having been covered with hides, and where local folklore recalled the era of Danish invasions (chapter 9). The 'Dane-skin' legend became so deeply embedded that local histories and guidebooks repeated it unquestioningly. However, the true number of doors in medieval England that were hide-covered probably ran at least into hundreds. Door-covering was practised throughout Europe in the Middle Ages, but most of the physical evidence has disappeared as a result of natural decay.

An introduction to the subject of 'Dane-skins' was published by Geddes, but the doors in question and their skins have never received collective study.[8] In the light of recent scientific advances, and the application of modern archaeological techniques, a review of the material evidence provided by the doors, their coverings and ferramenta is now feasible.

## Church doors with coverings of hide

Although only five examples of skin formerly attached to church doors are currently known to survive in Britain, the practice of covering the external faces of doors with hide was neither unusual nor geographically restricted. Recorded occurrences are found as far afield as North Yorkshire, Merionethshire, Herefordshire, Hertfordshire, London, Essex and Kent. These doubtless represent no more than the tip of the iceberg, and careful study of early doors with attached ferramenta may reveal further examples of skin trapped between the boards and iron fittings.

### Rochester Cathedral, Kent
Samuel Pepys was the first to record the presence of skins on church doors. He visited Rochester Cathedral on 10 April 1661, 'observing the great doors of the Church, as they say, covered with the skins of Danes'.[9] He was referring to the medieval west doors that no longer exist.[10] The portal dates from the first half of the 12th century and the arch is filled with a Romanesque stone tympanum that may be a slightly later insertion.[11] The present square-topped leaves date from 1888. They are heavily decorated with ironwork that does not replicate what went before.[12] No samples of the skin are known to have survived.

### Worcester Cathedral, Worcestershire
The remains of a skin covering were noted in the 1780s on the medieval north nave doors, and samples were collected.[13] Dr Peter Prattinton recorded: 'I recollect, when a Boy at School, between 1780–90, being shown what they said was a Human Skin on the inside of the N. doors of the Cathedral of Worcester'. The reference to the skin being on the 'inside' is noteworthy. Hides can only be attached to a flat surface, and the inner faces of virtually all medieval church doors are not flat: they are encumbered by ledges or framing.[14] Conditions at Worcester for skin preservation are good since the portal is protected by a deep, 14th-century porch.

The original north doors were removed to the crypt around 1800, and their replacements were again renewed in 1865–70 by Abraham Perkins (Cathedral Architect, 1848–73).[15] He retrieved the three surviving samples of hide from the doors that are now in the cathedral library. In the 1980s there was still a fragment of a major door exhibited in the crypt, the boards of which were jointed with loose tongues, a technique that Geddes assigned to the 11th and 12th centuries.[16] In 1894, W.S. Brassington, a local historian, made an interesting observation about the doors:

*Figure 2: Worcester Cathedral. Upper part of a 12th-century oak leaf from a former north nave door. Interior face, with a bevelled ledge fixed to the boards with clasping roves. A, line of clenched nails holding a hinge-strap on the exterior. B, diagonal black painted stripe on the ledge, remaining from 'barber's pole' decoration. C, medium grey painted band below the ledge. Fiona Keith-Lucas, courtesy of the Dean and Chapter of Worcester*

They are massive and unadorned, except by plain bands of iron, and bear traces of having been altered after they were made. The curved top of the doors corresponds with the arch-line of the old west doorway, which was Norman work of the 12th century. In the 14th century the great north porch was built, and then, probably, the old west doors were removed to the new porch.[17]

In 2025, the upper half of one door leaf was rediscovered in storage.[18] It comprises five oak boards of various widths, butt-jointed and secured with loose tenons (Fig. 2). On the front are the remains of two iron strap-hinges, one being original. On the back is a single ledge with bevelled angles, attached to the boards by nails, with lozenge-shaped clasping roves, used as washers under the nail-heads. The points of the roves were driven into the bevels. Remarkably, sandwiched between the ledge and the boards is a strip of hide, a feature that has not been recorded on any other Norman church door (Fig. 3). There is thus no doubt that the internal face of this door was covered with hide during construction, before the ledges were fitted. No evidence for an external hide covering can be detected.

The interior of the door was painted bright red on both the hide and the ledges. Additionally, there is a diagonal black stripe painted on the flat face of the extant ledge, indicating that it bore decoration, probably representing a 'barber's pole' design (Fig. 2).

Prattinton acquired a piece of the skin which he bequeathed to the Society of Antiquaries of London, *c.* 1840. It was described as:

> A portion of skin, supposed to be human, according to the tradition that a man, who had stolen the *sanctus*-bell from the high-altar in Worcester cathedral, had been flayed, and his skin affixed to the north doors, as a punishment for such sacrilege. The doors having been removed, are now to be seen in the crypt of the cathedral, and small fragments of skin may still be seen beneath the iron-work with which they are strengthened.[19]

Regrettably, the sample was taken from the Society's headquarters in Burlington House, and was on its way to be scientifically investigated, in an attempt to establish whether the skin really was human, when it was destroyed by German bombing during the Second World War. The Royal College of Surgeons holds a microscope slide of a hair taken from the skin.[20]

Three samples of skin are preserved in the cathedral library (Fig. 4). In 1904, it was noted that 'At Worcester Cathedral there is a large "slab" of human skin, that of another Dane who was caught in the act of stealing the *sanctus* bell.'[21] This does not entirely agree with Brassington's account of 1894, which lists four pieces of skin from Worcester: (i) in the museum of the College of Surgeons; (ii) in the museum at Audley End House (Ess.), along with a piece from Hadstock; (iii) in the chapter house at Worcester; (iv) in the possession of J. Noake, JP, of Worcester.

*Figure 3: Worcester Cathedral. North nave door leaf. Detail of the cut end of the bevelled ledge, showing the layer of hide trapped between it and the boards. Fiona Keith-Lucas, courtesy of the Dean and Chapter of Worcester*

The last was examined by Brassington, who described it thus: 'The relic might be mistaken for a piece of an old boot; it is of a rusty brown colour, and the man from whose body it came must have been an exceedingly thick-skinned person.'[22]

Today, the three fragments of hide are all glued to a sheet of Perspex and housed in a display case, with an attached antiquarian description: two pieces are thick, coarse and brown in colour (recalling Brassington's analogy of 'an old boot'); the third is pale cream and thinner. These fragments are seemingly from two different sources, and minute specks of red paint are discernible (Fig. 5). The display is accompanied by a legend that claims the skin to be derived from a flayed Dane, and records that the samples were taken from the north doors by Abraham Perkins.

### Westminster Abbey (1): St Faith's Chapel (sacristy)
Westminster Abbey was provided with two sacristies during the rebuild instigated by Henry III in 1245. The great sacristy was a large structure that lay in the angle between the north transept and the nave. The lesser sacristy, where the most valuable items were kept, was smaller, stone vaulted and highly secure; it was entered from the south transept (Figs 6, 7 and 19, door 1).[23]

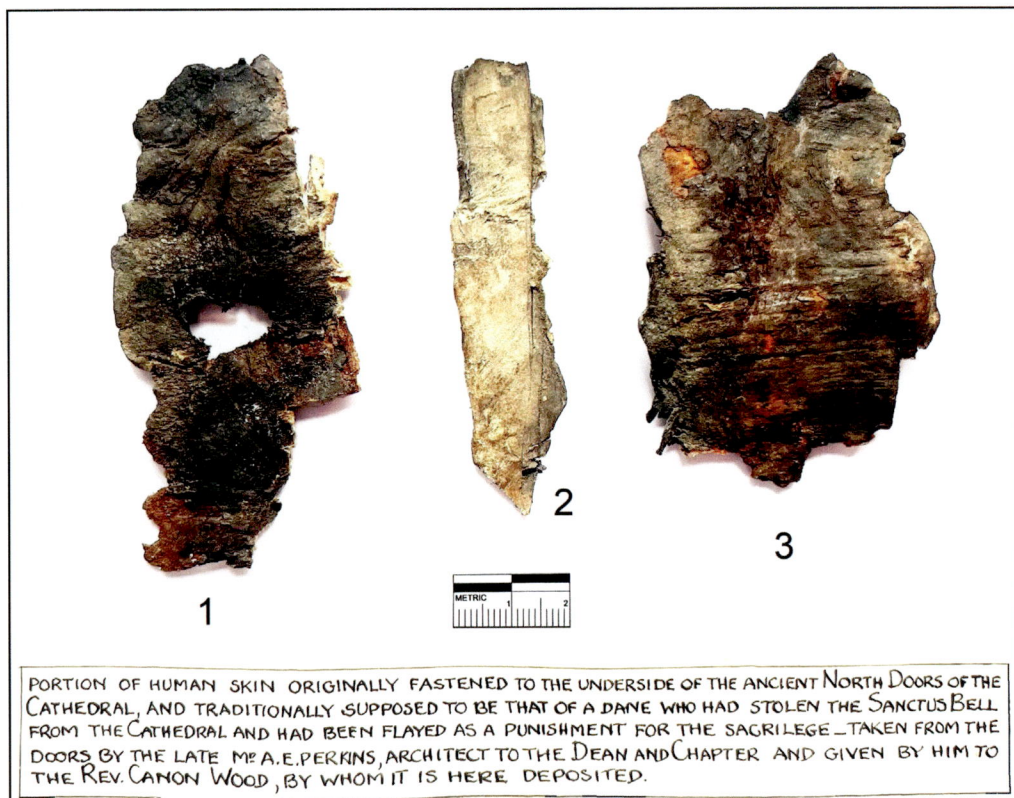

PORTION OF HUMAN SKIN ORIGINALLY FASTENED TO THE UNDERSIDE OF THE ANCIENT NORTH DOORS OF THE CATHEDRAL, AND TRADITIONALLY SUPPOSED TO BE THAT OF A DANE WHO HAD STOLEN THE SANCTUS BELL FROM THE CATHEDRAL AND HAD BEEN FLAYED AS A PUNISHMENT FOR THE SACRILEGE_TAKEN FROM THE DOORS BY THE LATE MR A.E.PERKINS, ARCHITECT TO THE DEAN AND CHAPTER AND GIVEN BY HIM TO THE REV. CANON WOOD, BY WHOM IT IS HERE DEPOSITED.

*Figure 4: Worcester Cathedral. Samples of hide taken from the north nave doors in the 19th century. Fiona Keith-Lucas, courtesy of the Dean and Chapter of Worcester*

In 1723, the historian John Dart briefly described the 'old revestry' (*i.e.* sacristy).

> This Revestry ... is inclosed with three Doors, the inner cancellated; the middle, which is very thick, lin'd with Skins like Parchment, and driven full of Nails. These Skins they, by Tradition, tell us, were some Skins of the *Danes*, tann'd, and given here as a Memorial of our Delivery from them. The Doors are very strong; but were notwithstanding broken open lately, and the Place robb'd.[24]

In 1812, another abbey historian, Rudolph Ackermann, repeated Dart's description, adding:

> Besides the grated door which now remains, which is of a massive construction, there was another without it, as appears from the marks of its bolts and bars. The latter was of great thickness and lay, when undrawn, in the substance of the wall: at the end of it was a hasp, secured by a lock and bolts on the opposite side.[25]

These accounts appear to describe the entrance to the sacristy as being fitted with three doors, two of which had disappeared before 1812, leaving only the inner one.

The use of the term 'grated' to describe the door implies that it incorporated iron grille-work, as the present door does. In 1853, the antiquary Albert Way reported that although the triple doors no longer existed, the masonry of the doorway 'preserves the indication of such threefold defence of a portion of the conventual church'.[26]

The accounts are inconsistent and confusing; there were never three doors in the opening. Since Dart described the innermost door as 'cancellated', he was presumably referring to a boarded door, backed by a latticed framework of battens, not its 'grated' successor.[27] In front of it lay the thick middle door, doubtless also comprising vertical boards attached to a heavy timber framework. That door was evidently covered on its outer face with hide, and had three strap-hinges. Dart's description of the skin as being 'driven full of nails' was an allusion to evidence remaining from a former display of ornamental ironwork.[28] This door is not to be confused with the Abbey's second skin-covered door in the chapter house vestibule, discussed in chapters 3 and 4.[29]

*Figure 5: Worcester Cathedral. Detail of one of the pieces of hide in Figure 4, showing a tiny patch of red paint. The larger patches of orange colour are rust-staining from ironwork that overlay the skin. Fiona Keith-Lucas, courtesy of the Dean and Chapter of Worcester*

Ackermann's description of the lost middle door as being 'of great thickness and lay, when undrawn, in the substance of the wall' is clarified by the extant archaeological evidence. The door was set in a rebated jamb in the middle of the wall. It was hinged on the western edge, opened outwards into the transept, and the three pintles remain in the jamb. The door was secured in two ways. First, there were bolts at mid-height on its east and west edges, the iron keeps for which are set in lead in the jambs. Second, a square-section, timber draw-bar was housed in a deep, oak-lined pocket in the core of the transept wall. The bar could be drawn out from the eastern jamb, pulled across the face of the door, and its free end housed in a shallow pocket in the western jamb; the hasp for a lock to secure the bar in the 'closed' position remains. References to the door being of 'great thickness' are exaggerated, since the evidence of the hinge-pintles, bolts and draw-bar all confirm that it was only 5 cm thick.[30]

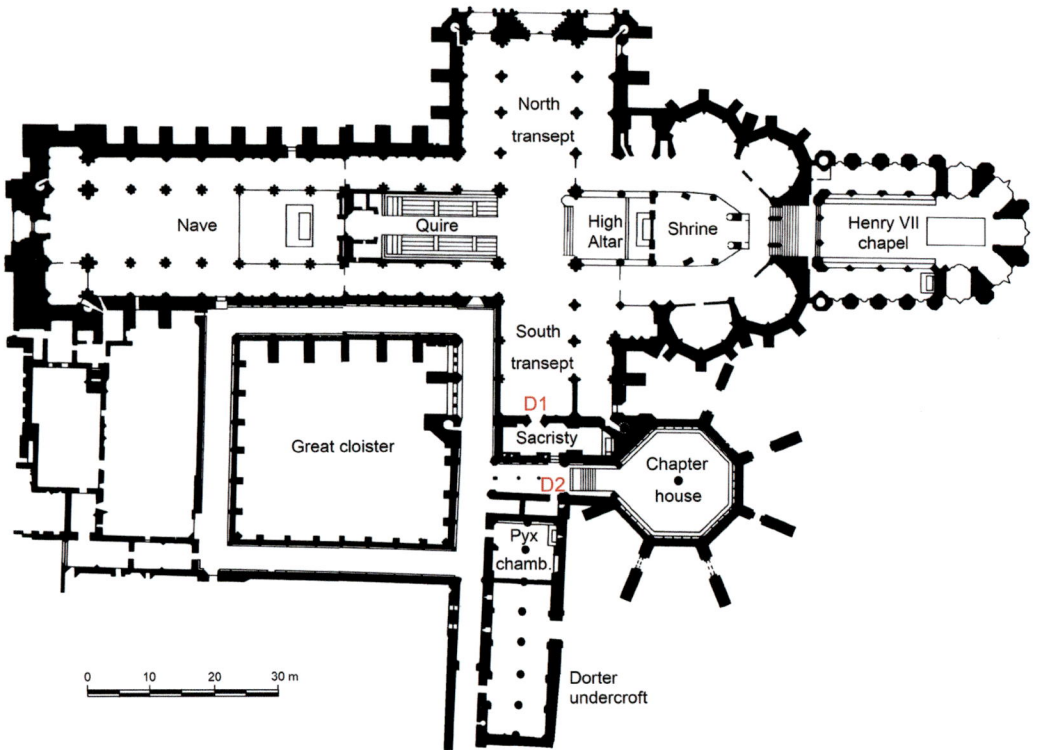

*Figure 6: Westminster Abbey. Location plan of the two skin-covered doors. D1, entrance to the sacristy from the south transept (original door no longer extant). D2, entrance to the former cellar, opening off the chapter house vestibule. Author, © Dean and Chapter of Westminster*

No description of the outermost door is recorded, and there is no fixing evidence for it in the masonry.[31] None of the doors seen by Dart in 1723 survive today, and the earliest views of the transept depict the present 'grated' door, described and illustrated by Ackermann in 1812 (Figs 7 and 8). It dates from the later 18th century.[32] The upper register emulates the form of a Perpendicular window of five main lights, with traceries; each light is 'grated' with a single iron stanchion.[33]

The sequence of doors to the sacristy is archaeologically reconstructible from the confused descriptions. In the 1250s there was only one door, set in a rebate in the stone reveal: it opened into the transept. The primary fixings all survive, as described above. The door would have comprised vertical boarding, which may have been held together by a series of horizontal ledges, or mounted on a heavy frame. The external (north) face was covered with animal hide and bore decorative ironwork. The inner door, recorded by Dart, must have been a later addition, fitted within the reveals of the opening, but not set in a rebate. It would have opened into the sacristy, to avoid collision with the outer door. The inner one was evidently boarded and mounted on

*Figure 7: Westminster Abbey. Interior of the south transept, showing the sacristy doorway (D1) in the south wall; a 13th-century hide-covered door formerly hung here. Ackermann 1812, vol. 2, pl. 27*

a frame of 'portcullis' or latticed construction ('cancellated'). No evidence is visible for the hanging of the inner door, that now being concealed by the large box-frame of its 18th-century successor. Increasing the security of the sacristy probably took place in the late 13th or 14th century, when the entrances to treasuries commonly comprised one inward-opening door and another outward-opening: *cf.* the early 14th-century Pyx Chamber doors of the Abbey and the 13th-century crypt doors at Wells Cathedral (Figs 9 and 133).

During one of the restoration campaigns in the late 18th century, both the sacristy doors were removed, and the present 'grated' one substituted.[34] In 1860, George Gilbert Scott commented: 'of these doors only one now remains'.[35] In 1868, Dean

*Figure 8: Westminster Abbey. Entrance to the sacristy, with the late 18th-century door that replaced the 13th-century one. The rebates, hinge-pintles and bolt-holes for the medieval door are visible in the jambs. Author, © Dean and Chapter of Westminster*

Stanley confusingly referred to the 'terrible lining ... affixed to the door of the Sacristy in the South Transept', as though the hide covering were still in existence.[36] Being fully visible from the interior of the church, the primary, hide-covered door may have been painted and gilded as part of the decoration of the transept. No samples of the skin are known to survive, or to have been submitted to a specialist for examination.

### Westminster Abbey (2): chapter house vestibule

For a detailed description of the door, see chapter 3 (Figs 21 and 22).

Scott discovered the door hanging in the south wall of the outer vestibule of the chapter house, probably first setting eyes on it in 1849, the year of his appointment as Surveyor of the Fabric to the Dean and Chapter. In 1852, he supervised the restoration of the vestibule to the chapter house, and in the following year wrote to Robert Marsh (Receiver General):

> I do not know whether you are aware of a strange archaeological phenomenon in an old cellar belonging to Mr Jell [Gell] which opens in the corner [of the vestibule] close to the Chapter House door from the Cloister. The curiosity consists of *skin* fastened to the door under the hinges, which skin has been proved to be *human*! This discovery was made by me with two others a year or more back.[37]

Scott's major architectural contribution to the fabric of the Abbey was the restoration of Henry III's octagonal chapter house which, since the Dissolution, had served as the repository for secular records. Work began on the new Public Record Office in 1851, and within three years the chapter house was being emptied of its contents. Restoration proceeded thereafter, but Scott had already been carrying out preparatory investigations since his appointment.[38]

Post-medieval reconfiguring of the vestibules had seriously changed their 13th-century layout and appearance and, moreover, the stone stairway that led from the cloister up to the library (formerly the monastic dormitory) had been blocked and

*Figure 9: Westminster Abbey, east cloister. Double doors opening into the Pyx Chamber (treasury). The outer door is backed by a diagonally latticed framework, and the inner door by one of rectilinear ('portcullis') form. To the left is also glimpsed the door to the monastic day-stair leading to the dormitory (now the library). Author, © Dean and Chapter of Westminster*

infilled from above. Beneath the low soffit of the stairway was a small, oddly shaped storeroom with no designated function; often described as a 'cellar', it was accessed from the outer vestibule via a low doorway in which hung a battered old oak door (Fig. 10). Being a part of the Abbey that was not in ecclesiastical use, or accessible to public view, the door went unnoticed: no references to it or the cellar occur prior to 1860.

It was only when Scott unlocked the door and entered the cellar that its significance began to dawn. The door clearly did not originate here, and had been cut down and reversed when it was hung: its original inner face was now towards the exterior (north). Both faces bore the last remnants of medieval ironwork and the tattered remains of the hide that had once covered the timber. Scott recorded his discovery:

On the inner side of the door I found hanging from beneath the hinges some pieces of white leather. They reminded me of the story of the skins of Danes, and a friend to whom I had shewn them sent a piece to Mr Quekett, of the College of Surgeons, who, I regret to say, pronounced it to be human. It is clear that the door was entirely covered with them, both within and without. I presume, therefore, that this, too, was a treasury; and I have a strong idea that it then formed a part of, and that its door was the entrance to, the Pyx Chamber, and it is possible that, after the robbery of the chamber [1303] ... the King, finding that the terror of human skins offered no security, remodelled the chamber, and intrusted the safety of his treasury to the less offensive, but more prosaic, defence of massive and double doors and multitudinous locks.[39]

The revelation regarding the source of the skin was made in the late 1850s, before Quekett died (p. 26). The door was also seen in the 1860s by Henry Harrod, who added an interesting detail: 'Inside and outside of the door by which this passage is entered is nailed the skin of a fair-haired ruddy-complexioned man.'[40] In 1886, a visitor found 'portions of the skin still remaining on an old disused door in the corridor by the entrance to the chapter house'.[41] Parts of the hide still remain attached to the door, trapped by ironwork (Fig. 30). There are also small samples preserved in the Westminster Abbey Collection (Fig. 32). The fate of the sample in the museum at Audley End House is unrecorded.

Most 20th-century guide-books to the chapter house and Pyx Chamber made no mention of the antiquity of the door, commenting only on the hide, for example: 'The thirteenth-century door was covered with human skin, of which fragments remained until a few years ago; it was put there as a warning to thieves.'[42]

*Figure 10: Westminster Abbey. Chapter house, outer vestibule. Reused door, formerly hide covered, opening into a small cellar. Author, © Dean and Chapter of Westminster*

### St Botolph's church, Hadstock, Essex

For descriptions of the north and west doors, see chapter 5 (frontispiece, Figs 11 and 89).

The north nave door is the most notorious of all the English church doors for the preservation and study of its fragmentary skin covering. Over the course of the last three centuries, it has been referenced on innumerable occasions

*Figure 11: Hadstock church, from the north-east. The north nave door, formerly hide covered, lies within the porch adjacent to the transept. Author*

in antiquarian, folklore and popular literature. It has invariably been reported as 'Dane-skin', and has been the object of several scientific examinations, details of which are given in chapter 2.

The antiquary William Stukeley visited eastern England in 1724/25 and heard about the Hadstock door: he wrote 'At Hadstok they talk of the skin of a Danish king nailed upon the church-doors'.[43] A description of the church by a visitor from Cambridge in 1746 states:

> On ye great North Door of ye Nave, according to Tradition, is nailed and fastened by curious Plates of Iron ye skin of a Dane. But thro' Length of years and Peoples Curiosity in cutting off Pieces from it very little remains of it, except what is covered by ye aforesaid Plates.[44]

In 1768, the Rev'd Philip Morant, Essex's principal historian of the 18th century, noted:

> The north door of the church is said to have been covered with the skin of a Danish King, nailed on with many hundreds of nails. Only small bits of the skin remain around the nails, which is extremely hard. – There was the like tradition about Copford church.[45]

Stukeley most likely introduced the romantic notion that the skin was that of a Danish king, and Morant unwisely embraced it. Perpetuated by several later historians, Peter Muilman's description (1770) contains additional details:

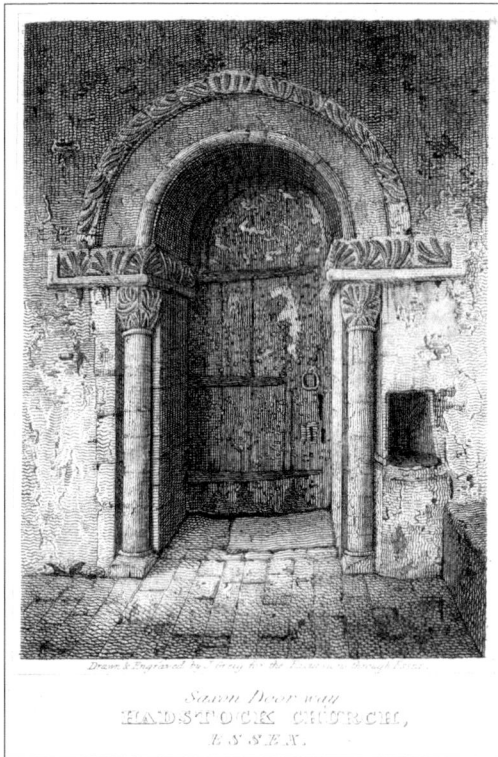

Figure 12: *Hadstock church. Drawing of the north doorway in 1818, published in Cromwell's* Excursions in the County of Essex. *Drawn and engraved by J. Greig*

The north door of this church is much adorned by thick bars of iron work, of an irregular form, beneath which is a sort of skin, said to be that of a Danish king {see Dr Stukeley's *Itinerarium Curiosum. p. 75*}. It is nailed on with large nails.

This is followed by a footnote regarding Copford church door, copied from Richard Newcourt's *Repertorium* of 1710 (see below), to which Muilman added a final sentence about Hadstock:

Of the like consistency is the skin on this church door; which notwithstanding the number of years it hath been there, does not appear to be much decayed: nor has the rust of the iron with which it is covered, scarcely injured it.

The description is clearly earlier, and was copied from an unacknowledged source, when the skin was better preserved. Like the note about Copford, it too may have originated with Newcourt, or an earlier source. Curiously, Newcourt's published entry for Hadstock is so brief that it makes no mention of the church, let alone the door; this must surely have been an accidental omission. Maybe Muilman had sight of a manuscript, or hand-written annotations in a printed volume of the *Repertorium*.

Thomas Cromwell, in his popular *Excursions in the County of Essex* (1818–19), observed

> The north door of this church is much adorned with thick iron-work of an irregular form, and underneath is a sort of skin, said to be that of a Danish king; it is nailed on with large nails. There is a tradition of a similar kind respecting the church door at Copford, which is also covered by a skin defended by iron-work.[46]

Exceptionally, Cromwell also included an engraving of the 'Saxon doorway', the first published illustration of a skin-covered door. It shows three strap-hinges and small patches of hide still adhering (Fig. 12). A pencil sketch made in 1809 depicts the doorway rather more crudely. The hinges are shown, together with two concentric iron hoops in the arch (Fig. 72).[47]

In Thomas Wright's history (1835), we read:

> formerly, what was traditionally said to be a Danish human skin, was nailed against the door here, and covered with iron rib-work: the iron remains, but the skin has been taken away by degrees, and only a small piece of it has been preserved at the parsonage, which from its appearance tends strongly to confirm the traditionary account.[48]

William Coller, writing in 1861, gives the impression that he had personally inspected the church: 'a massive Norman arch forms the northern entrance'. The door was 'ornamented with ancient ironwork, beneath which was a skin of enormous thickness, which appeared to have been tanned; and this tradition represents as the skin of a Dane, who was flayed alive for sacrilege in this church'. He then cited Neville's comments (p. 85).[49] George Tyack was in error when he stated that 'a new door was erected in 1846 at Hadstock, to which the fragments of skin that still adhered to the old one were not transferred'.[50] The door was extensively repaired, but not entirely replaced.

In 1789 Sir Henry Englefield exhibited at a meeting of the Society of Antiquaries a fragment of iron that had been taken from the church door, and a portion of skin found underneath it.[51] When the door was removed for repair in 1846, a piece of the timber with ironwork and skin attached was acquired by Richard Neville, archaeologist and fourth Baron Braybrooke for his museum at Audley End House, Saffron Walden.[52] In 1904, another fragment of skin was exhibited to the Cambridge Antiquarian Society by Alderman Deck.[53] The fate of all these pieces is unknown. Yet another fragment was presented by Quekett to the Hunterian Museum, where it was destroyed during the Second World War. Finally, 'a square inch of "Dane's skin" from Hadstock church fetched the sum of three guineas' at an auction in London in 1905.[54]

Fortuitously, one fragment of skin, taken from the door in 1791, is preserved in Saffron Walden Museum, along with a piece of ironwork (Fig. 69).[55] The supposed human origin of the skin has frequently been rehearsed in popular literature, and in some scholarly works, down to the present day.

### St Michael's church, Copford, Essex

For descriptions of the door and five surviving skin fragments, see chapter 8 (Figs 101–103).

Morant, referring collectively to Hadstock and Copford, wrote: 'The story of the Danes skins nailed on the church doors, whether true or false, may be seen in Mr. Newcourt.'[56] In his *Repertorium* (1710), Newcourt wrote:

> The Doors of this Church, are much adorn'd with flourish'd Iron-Work, underneath which is a sort of Skin taken notice of in the Year 1690, when an old Man at *Colchester*, hearing *Copford* mentioned, said, that in his young time, he heard his Master say, that he had read in an old History, that the Church of *Copford* was robb'd by *Danes*, and their Skins nail'd to the Doors, upon which, some Gentlemen being curious, went hither and found a sort of tann'd Skin, thicker than Parchment, which is suppos'd to be human Skin, nailed to the Door of the said Church, underneath the said Iron-Work, some of which Skin is still to be seen.[57]

Muilman repeated this *verbatim* in his history of 1772,[58] and Wright's history in 1836 merely recited the 'Dane-skin' story:

> The doors are covered with ornamental flourishes of iron work, and under these may yet be seen the remains of a kind of tanned skins, thicker than parchment, which are traditionally recorded to have been the skins of Danes, who broke into and robbed this church.[59]

Cromwell (1818–19), reiterated the reference to 1690 and the 'Dane-skin' story, adding that their skins 'were nailed to the doors underneath the iron-work before mentioned, which it would appear had been added for the purpose of preserving this singular example of retribution'.[60] This is the earliest account that attempts to explain how the skins came to be trapped under the ferramenta. Like most writers who mention Copford, Cromwell alludes to the church's 'doors' in the plural (north and south nave).

One-hundred-and-fifty years after Newcourt, Coller loquaciously devoted a section in his history to 'The Dane's skin at Copford'.[61] Some of his text was borrowed directly from Newcourt.

> ... its old entrance door [presumably south] will tempt the traveller to turn towards the antique fabric. This door is ornamented with rude flourishes of rusty iron work, which formerly fastened securely to the wood beneath a thick substance outwardly resembling parchment – similar to that at the church at Hadstock. Tradition, which takes maternal charge of many a marvellous tale, connects this leather-like and shrivelled coating with the system of savage retribution found in the code of justice in the olden time, but happily blotted from its pages in the present century. Some Danes, saith this authority, robbed the church – considered one of the most heinous of crimes in the mediaeval ages – and were subjected to the fearful process of flaying alive, their skins, carefully preserved, being thus affixed to the door as a terrible memento of the wretches who had dared to raise their sacrilegious hands against the house of God. The peculiar character of the door appears to have first attracted notice on the restoration of the church in 1690, when some beautiful ancient fresco paintings were found beneath the desecrating whitewash on the walls; and 'an old man at Colchester said that in his young time he heard his master say that he had read in an old history that the church of Copford was robbed by Danes, and their skins nailed to the doors.' This is the foundation of the tradition. Anxious to test it, we procured a piece of the skin, of which time and curious visitors have now left scarcely a shred. This we submitted to a scientific friend, skilled in anatomy, who, after softening and subjecting it to rigid examination, pronounced it to be 'part of the skin of a fair-haired human being' – thus confirming, to a considerable extent, the tale of torture which garrulous tradition has told to her wondering auditors.

### St Anne and St Laurence's church, Elmstead, Essex

For a description of the door, see chapter 8 (Figs 106–108).

The north doorway to the nave had long been blocked and was only reopened in 1935, when the Norman door was discovered.[62] It bore remnants of ornamental ironwork, beneath which were fragments of hide that had once covered the external face. Unlike all the other doors discussed here, this one and its skin covering had no previous recorded history. If there had been a 'Dane-skin' legend associated with the door, memory of it had faded.

## St Nicholas's church, Castle Hedingham, Essex

For descriptions of the doors, see chapter 8 (Figs 115–118).

The church possesses three 12th-century doors, all decorated with ironwork.[63] The south door is locally known as the 'skin door'. Tyack observed: 'Beneath the iron scroll-work on the Norman doors of the parish church of Castle Hedingham leather, like parchment in appearance, has been found, and the local legend says it is the skin of a man. The writer of this article is not aware that it has ever been thoroughly examined.'[64] No specific information has been gleaned regarding the possibility that samples of skin were retrieved from the doors, but there has been extensive replacement of some of the boards, and it is entirely feasible that trapped pieces of skin were discovered during repair work. No samples of the hide are known to exist.

## St Mary's church, Little (East) Thurrock, Essex

The church was heavily restored, 1878–84. The Norman nave has opposing doorways: the southern is 12th century, and the northern is now largely modern but includes 14th-century work.[65] William Quekett, rector of Warrington (Lancs., now Ches.), recalled that a piece of skin found during restoration work at East Thurrock church was sent, unsolicited, to his brother, John. William was with John at the College of Surgeons when a letter arrived, containing an enclosure 'which looked like part of the bottom of an old shoe, of the thickness of half-a-crown, of dark colour, elastic, and with the markings of wood upon it'.[66]

> The letter was from a churchwarden of East Thurrock, in Essex, who wanted the professor to tell him, if possible, what the substance was, without having any particulars of its history. Having washed it, and cut a thin slice, he discovered under the microscope that it had all the structure of human skin, and on a more minute examination that it was the 'skin of a light haired man, having hair of a sandy colour'. He wrote to the churchwarden telling him of the result of his examination. The latter replied that he (the professor) had 'proved the truth of a great tradition which had existed for years in East Thurrock'.
>
> The churchwarden went on to say that 'on the west door of the church there had been for ages an iron plate of a foot square, under which they said was the skin of a man who had come up the river [Thames] and robbed the church. The people had flayed him alive and bolted his skin under an iron plate on the church door as a terror to all other marauders. At the restoration of the church, which was then going on, this door had been removed, and hence he had been able to send the specimen.'
>
> It appears to have been assumed that the marauder who had been skinned was a Dane. Mr W. Quekett had a bit of the skin fixed as a specimen for the microscope, and wrote on the slide, 'This is the skin of a Dane, who, with many others, came up the river Thames and pillaged churches. Caught in the act at East Thurrock, Essex, and flayed alive.'

> The fate of the specimen on the slide is interesting. Mr Quekett lost it, and knew nothing for many years of what had become of it. In or about 1884, apparently, he was reading aloud to some gentlemen in the hall of the 'Palace Hotel', Buxton, an account of a meeting of the British Association at Penzance. In this account he came across the fact that at the meeting a microscopic object, among others of special interest, had been exhibited by a gentleman in the neighbourhood, viz., a 'Dane's skin' and that the specimen at Penzance had on it, word for word, what he had written on his lost treasure.
>
> He exclaimed 'Why this is my Dane's skin! I lost it twenty years ago.' After telling those present how he had obtained the specimen, he said aloud, 'I wonder who that man is'. Immediately afterwards the porter, who had heard the conversation, said, 'Please, Mr Quekett, I can tell you who that gentleman is; I was his footman and valet for four years; it is Mr -------, who lives at ------- Castle, near Penzance.' Mr Quekett wrote at once to the gentleman, whose name he does not give, claiming the specimen, and asking him how he had come in possession of it. The gentleman replied that the description of the specimen and the account of the inscription were perfectly correct; that it had been given to him by a lady in London; that he greatly valued it; and that should Mr Quekett ever be in his part of the country, and should wish to see it, he would have great pleasure in showing it to him.[67] *Beati possidentes.*[68]

The fragment sent to Quekett cannot have been the whole of what was found under the iron plate, but there is no further information recorded about this occurrence of skin on the church door: presumably the remainder has been lost.[69]

### St Helen's church, Stillingfleet, North Yorkshire
The south door and its elaborate Norman ironwork have been intensively studied and published (Fig. 13).[70] The former presence of skin on the outer face was noted in the 1880s, but is now lost.[71]

### St Mary's church, Pembridge, Herefordshire
The church is largely of the 13th and 14th centuries, but with some evidence of Norman origins. The north door is 14th century and has ornamental ironwork, including a fine sanctuary ring.[72] In 1901, the Rev'd J.B. Hewitt noted:

> Behind the ironwork of the knocker is a leathery substance – much decayed and covered with paint, but suggesting the awful fate reserved for Danes and those caught red-handed in sacrilege. A portion of this has been submitted for examination under the microscope by an expert, who expresses a strong opinion that it is human. It may, therefore, be fairly added to the known instances of human skin upon church doors.
>
> It will be seen that the skin remains perfect in only one of the four divisions of the ring.[73]

The description invites the suggestion that the unnamed expert was John Quekett, and if so, it would indicate that the examination took place prior to his death in 1861. Later commentators assumed that the skin was lost, but a small fragment has fortuitously survived and is now in Hereford Museum (Fig. 14). It is c. 2 cm square and very dark in colour.[74]

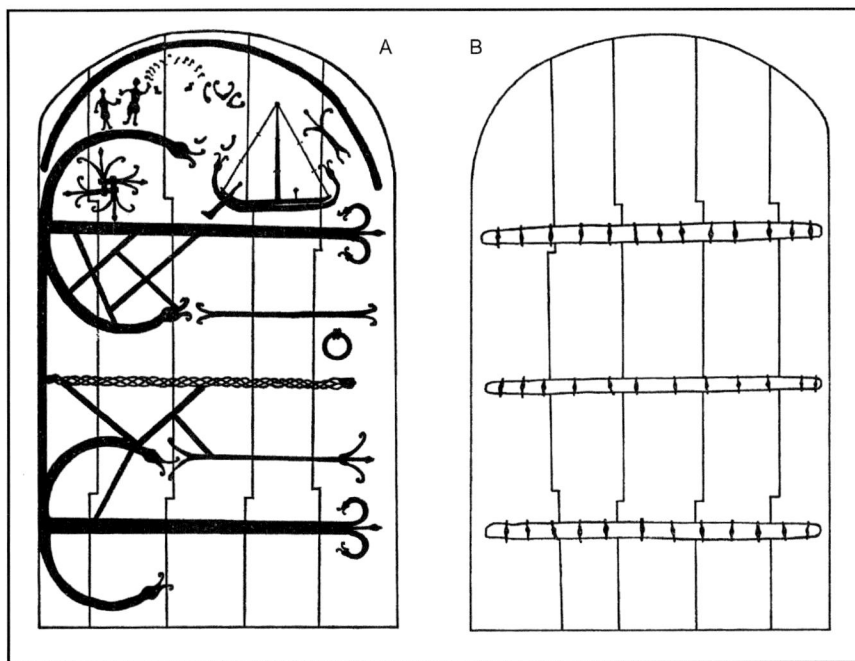

Figure 13: Stillingfleet church, south door. Reconstruction drawings of both faces. The decorative ironwork on the exterior was originally set on a background of hide that would probably have been painted red. Traces of the hide were still present under the Viking ship in the late 19th century. Addyman and Goodall 1979, fig. 19

### Other possible occurrences

Vague mentions of 'Dane-skins' elsewhere on church doors in England and Wales are uncorroborated: *e.g.* at Bosham (W. Suss.) and Stogursey (Som.).[75] The west door at Piddletrenthide (Dor.) is said to have borne skin too.[76] At the coastal church of Llanaber (Gwynedd) in West Wales, remains of hide on the door were first noticed before 1908,[77] and it was reported again in 1949 by Cledwyn (Lord) Hughes.[78]

Leather-covered doors are to be expected in medieval secular buildings, but they have not attracted scholarly attention. A 15th-century example was found during the restoration of Hertford Castle in 1968–71.[79]

Figure 14: Pembridge church, north door. Fragment of hide recovered from under the sanctuary ring; now in Hereford Museum. © Herefordshire Museums & Galleries

# 2

# Antiquarian study of hide-covered doors

## The preoccupation with Danes and flaying

The first attempt to list and discuss church doors with attached skins was by Albert Way (1805–74), a prominent Fellow and sometime Director of the Society of Antiquaries of London. He had wide connections and antiquarian interests, and was fascinated by the legends associated with the fragments of skin attached to medieval church doors. He researched all the cases that came to his attention, and published a paper on human flaying in 1848.[1] His interest in the subject appears to have been awakened by a visit to Hadstock in about 1840, where he was informed that the skin trapped beneath the hinges on the church's north door was that of a sacrilegious Dane who had been flayed as punishment for his actions.[2]

Archaeological evidence confirms that some church doors in Britain were clad with hides from at least the mid-11th century to the late 13th century. The earliest are the Westminster vestibule and Hadstock skins, and the latest are those at the Westminster sacristy (mid-13th century) and Pembridge church (potentially 14th century). In the mid-12th century, Theophilus, a monk and artisan, described the construction of church doors, and their covering with animal hide (p. 148). As already noted, the first recorded observation of skins on British doors was by Pepys, who saw them at Rochester in 1661. He was also the first person to record the belief that the skins were of human origin: 'as they say, covered with the skins of Danes'.[3] The earliest mention of 'Dane-skins' at Copford church was in 1690 (p. 15).

Thereafter, references to the supposed human origin of the hides became popular with historians and romanticists, and 'sacrilegious Danes' were most commonly cited as the unwilling donors. In the case of Hadstock, the victim was claimed to be an unnamed Danish king (p. 13). The majority of later writers accepted, unquestioningly, that the skins were not only human, but also Danish, notwithstanding the fact that there was a serious chronological discrepancy – amounting to several centuries, in some instances – between the era of the pagan Viking invasions and the construction of most of the doors that bore skin coverings. Some writers ignored this inconvenience, but others tried to bridge the chasm by arguing that the doors had been reclaimed from earlier churches. However, in those instances where the design and decoration of the church doors were indisputably well after the Conquest, as at Worcester and Pembridge, Viking-adherents claimed that the skins themselves might have been recycled on new doors: fantasy appeared to have no limits.

In the case of Westminster (2), two different origins for the skin were championed. On the one hand, it was regarded as a 'Dane-skin' by Gilbert Scott, and on the other it was surmised by Henry Harrod to be the product of a much later flaying, consequent upon theft and sacrilege within the monastic community in 1303. Scott argued that the door came from the original 11th-century entrance to the Pyx Chamber, which lies next to the vestibule and is part of the monastic dormitory undercroft (Fig. 19). If so, we do not know where that doorway was located, the appellation 'Pyx Chamber' having not yet been coined. The Trial of the Pyx did not take place there until the 14th century, and it was only in the 19th century that the term 'Pyx door' was invented, despite there being no basis for such an assertion. The misnomer has continued in use until the present day.[4]

Scott argued that when the Pyx Chamber was robbed of the royal treasure, the old, weak door in the vestibule was abandoned and superseded by the two heavy doors with portcullis framing that still exist in the east cloister walk (Fig. 9). The instigator of the burglary was a merchant, Richard de Pudlicote,[5] who achieved the heist in collaboration with several monks, including the sacrist of Westminster Abbey. Pudlicote and other non-monastic culprits were apprehended and hanged, but it has been hypothesized that the instigator was flayed and his skin nailed to the old door.

In 1868, Dean Stanley, whose writing is not always notable for accuracy, unequivocally linked the skin on the vestibule door to the 1303 burglary:

> Inside and outside of the door ... is nailed the skin of a fair-haired, ruddy complexioned man. The same terrible lining is also affixed to the door of the Sacristy in the South Transept of the Abbey. These fragments of human skin are generally said to belong to a Dane; but, in fact, there is no period to which they can be so naturally referred as to this [*i.e.* 1303]; and they doubtless conveyed the same reminder to the clergy who paced the cloisters or mounted [the steps] to the dormitory door, as the seat on which the Persian judges sate, formed out of the skin of their unjust predecessor, with the inscription, 'Remember whereupon thou sittest.'[6]

Two years later, Harrod, in a paper read before the Society of Antiquaries, repeated Stanley's assertion, *verbatim*.[7] However, both omitted to mention that there is no evidence to support the hypothesis that Pudlicote or any of his accomplices were flayed. Nor was any attempt made to explain the origin of the skin on the sacristy door in the south transept.[8]

## Early scientific attempts to identify the animal species of hides

With the rapid advancement of scientific research in the 19th century, a desire emerged to seek irrefutable proof that the door skins were indeed human. Macabre instincts needed to the satisfied. The instigator of a series of investigations was Albert Way, whose first foray into scientific research was in relation to the Worcester skin. At his request, further small samples of skin were retrieved from the doors and in 1847 they were submitted to John Quekett, who was then Assistant Conservator (later Professor of Histology) at the Museum of the Royal College of Surgeons. On the Worcester skin, he reported as follows:

> I am perfectly satisfied that it is human skin, taken from some part of the body of a light-haired person, where little hair grows ... and two hairs which grow on it I find to be human hairs. The hairs of the human subject differ greatly from those of any other mammalian animal, and the examination of a hair alone would have enabled me to form a conclusion.[9]

Interestingly, Way records that Jabez Allies, the antiquary who procured the sample, 'was decidedly of opinion that the skin had been laid upon the wooden leaves of the door, at the time of its original construction'. He seems to have been the first person to state this important archaeological fact. Although most other commentators alluded to the skins being preserved *under* iron hinge-straps, they passed no comment.

Encouraged by this result, Way lost no time in procuring a fragment of the skin from Hadstock, in order to subject it to a similar test. He obtained a sample from Richard Neville (later Lord Braybrooke), and submitted it to Quekett in July 1847, who responded favourably: 'I have been again fortunate in making out the specimen of skin you last sent me to be human' (p. 28). Three hairs were regarded as the clinching evidence. Quekett published his findings, accompanied by a lithograph that purported to demonstrate how the structure of hairs matched a modern human sample (Fig. 15).[10] Although he illustrated three 'ancient' examples of

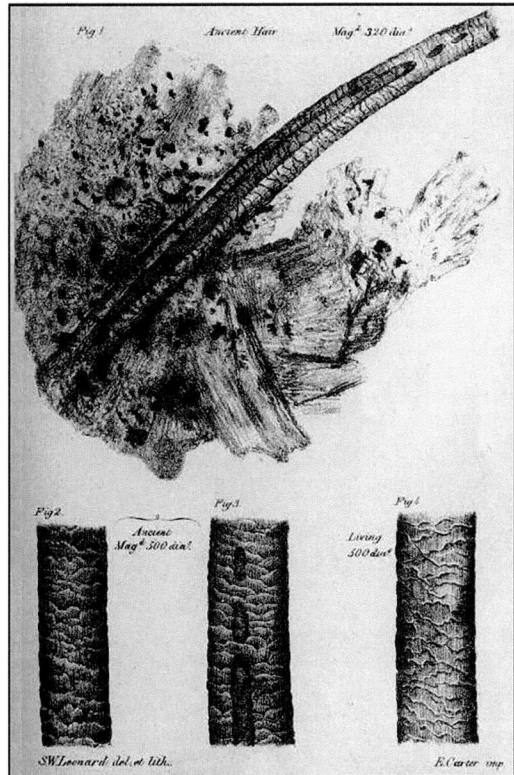

*Figure 15: Drawings of human hair made from microscope slides, to illustrate John Quekett's paper of 1848. 1, ancient hair (×320 mag.); 2, 3, ancient hair (×500); 4, living hair (×500). Quekett 1849, pl. 24, figs 1–4*

supposedly human hair, he did not state from which church door skins they were taken. The information was probably recorded on the original slides, but they were destroyed during the Second World War.[11] Apparently, Charles Townley, rector of Hadstock, 1838–70, also instigated some form of scientific examination; it returned the same 'satisfactory' result, but no details are recorded.

Next, Way turned his attention to Copford church and his enquiry to the rector, Kenneth Bayley (1845–61), received the following response: 'There are no remains of skin on the door at the present time. I have however in my possession a short MS account of the parish, written during the incumbency of John Dane, 1689–1714.' He then recited the description published by Newcourt in 1710 (p. 15).

Bayley added the following information: 'Since writing the above I have heard that what remained of the skin was removed about four years ago. I hear, however, of two pieces in this neighbourhood, and if I can succeed in procuring either of them, I will forward it to you.' On the following day, the rector duly produced one of the fragments. It had been taken by a carpenter from underneath the ironwork, c. 1843, when the church was undergoing repair. The sample was submitted to Quekett, who pronounced it to be human, again on the basis of surviving hairs.[12] Way expressed great satisfaction at the results obtained:

> The value of natural science as a friendly ally to archaeology in supplying conclusive evidence on a question which must, without such aid, have been left to vague conjecture, has been strikingly shown in the present instance. The singular corroboration of the truth of popular tradition, thus undeniably established, may serve to remind us that no circumstance, however apparently trivial or absurd, is without utility in the investigation of the history and usages of ancient times.[13]

Bayley must have retained one of the pieces of skin, passing it on to his successor, Peter Wood (rector, 1861–78), who in turn donated it to Colchester Museum (p. 32). Somehow, the church acquired three more, framed samples of skin, which it still holds (Figs 16 and 104). Later, Quekett received a sample taken from the Westminster vestibule door by Gilbert Scott, probably via Way, and this too he pronounced to be human.[14] At an unrecorded date, Quekett was also sent a skin sample from Little Thurrock, but his report has not survived. Almost certainly the Pembridge sample was identified by Quekett too; and there may have been others of which we have no details.

As far as we know, all the skin samples submitted to Quekett were emphatically identified as human, based on a few hairs attached to each skin. This appeared to be so conclusive that few dared to question Quekett's assertions. Consequently, most antiquaries were now fully accepting of his conclusion, reinforcing both scholarly and macabre popular opinion concerning the origin of 'Dane-skins'. They became the subject of diverse articles.[15] Further investigations and pronouncements by scholars in the 20th century were severely biased as a result of Quekett's string of positive identifications. When Sir Arthur Keith re-catalogued the collection of the Royal College of Surgeons in 1923, he accepted Quekett's identifications unquestioningly.[16]

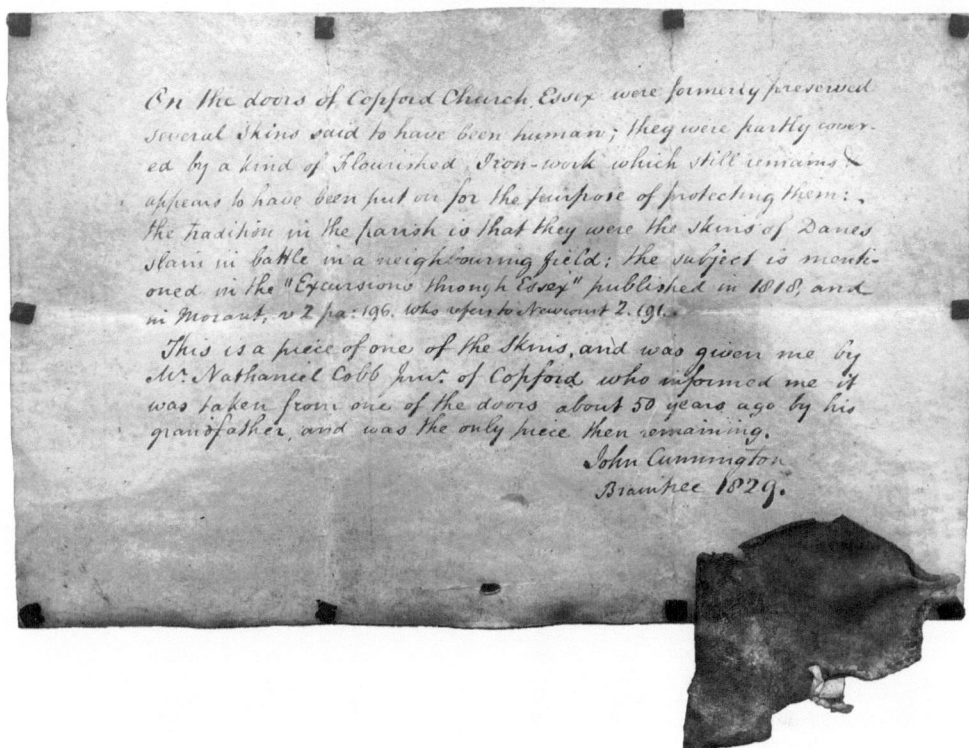

On the doors of Copford Church, Essex were formerly preserved several Skins said to have been human; they were partly covered by a kind of Flourished Iron-work which still remains & appears to have been put on for the purpose of protecting them: the tradition in the parish is that they were the skins of Danes slain in battle in a neighbouring field; the subject is mentioned in the "Excursions through Essex" published in 1818, and in Morant, v 2 pa: 196, Who refers to Newcourt 2. 191.

This is a piece of one of the Skins, and was given me by Mr Nathaniel Cobb Junr. of Copford who informed me it was taken from one of the doors about 50 years ago by his grandfather, and was the only piece then remaining.

John Cunnington
Braintree 1829.

*Figure 16: Copford church. Hide fragment (i), a crumpled piece attached to a note by John Cunnington, 1829. Author, courtesy of St Michael's Church*

In the following year, the RCHME briefly described the Westminster cellar and door: 'of plain battens with strap-hinges and a large iron staple and padlock, mediaeval; under iron hinge remains of skin, said to be human'.[17]

Responsibility for scientifically 'proving' that all the known skins on church doors were human thus rested with Quekett alone. So who was he, and what were his professional credentials?

### John Thomas Quekett, 1815–61 (Fig. 17)

John Quekett was the youngest of four brothers, born at Langport (Som.). At an early age, he developed a special interest in microscopy, not just for investigation but also in construction and use, and at the age of only sixteen he gave a course of lectures to the pupils of his school. He decided on a medical career, trained initially with a surgeon at Langport, and then came to London as apprentice to his brother Edwin, and was entered as a student at the London Hospital Medical College, and at King's College.

He qualified in 1840 and won a three-year Studentship in Human and Comparative Anatomy at the Royal College of Surgeons. During that time

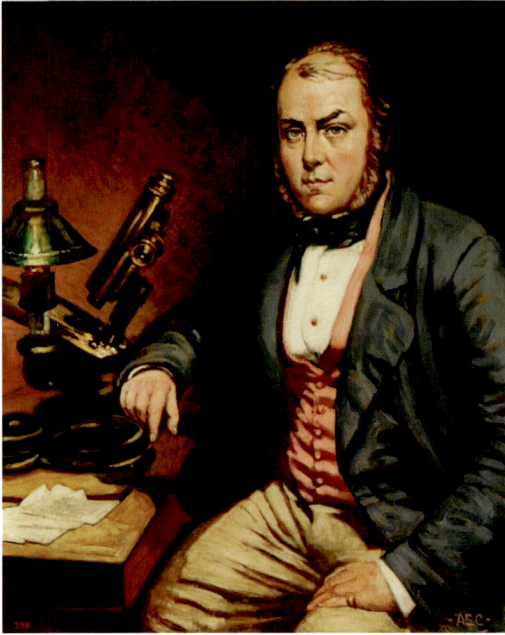

*Figure 17: John Thomas Quekett (1815–61). Portrait in oils by Alfred Egerton Cooper, after Elizabeth and William Walker. By kind permission of the Royal College of Surgeons of England*

he made some 2,500 microscopical preparations and the collection is still held by the College. His diaries refer to men he knew who were among the most prominent microscopists of the time. He wrote a *Practical Treatise on the Use of the Microscope* in 1848, and gave a lecture 'On the Value of the Microscope in the Determination of Minute Structures of a Doubtful Nature, as exemplified in the Identification of Human Skin attached many Centuries ago to the Doors of Churches'.[18] He was a popular lecturer and had two volumes published of *Lectures on Histology*, and he gave some instruction to Prince Albert on the use of his large silver microscope. He was appointed Assistant Conservator of the College's Hunterian Museum in 1843, and Conservator in 1856. Meanwhile, Quekett was appointed Professor in Histology at the College in 1852.

The Microscopical Society of London was founded in 1839 in the house of Quekett's brother, Edwin, and John was secretary of the society, 1841–60. In 1866 it received a royal charter, becoming the Royal Microscopical Society. A new society in London was later named the Quekett Microscopical Club in his honour. In 1857 Quekett was elected a Fellow of the Linnean Society, and was made a Fellow of the Royal Society in 1860. However, his health was failing and he requested by letter that the society should not proceed with the intent to elect him as president. The letter was unfortunately delayed and he later found that he had been elected. He gave a presidential address that year, but died six months later at Pangbourne (Berks.).

Specimens reached him from all over the world and he was consulted on numerous matters. When Quekett died in 1861, his large collection of samples and records was bequeathed to the Royal College of Surgeons, but tragically almost everything was destroyed by the bombing of London in 1941.[19] Tyack recorded, in 1898, that fragments of skin from Worcester, Hadstock and Copford were preserved in the College's Museum.[20]

A version of his paper read before the Microscopical Society of London on 26 April 1848 reveals how the subject of skin identification developed as a result of communications between Way and Quekett. The lecture essentially comprised reading a series of five verbose letters, and in his preamble, Quekett made the following observations (abridged):

> In any two animals, of totally different genera, the hair when examined chemically may be in composition identical, but when submitted to the microscopic test, may be found so manifestly different that the unpractised eye can readily discriminate between them ... [In] 1847, I was asked by Sir Benjamin Brodie,[21] whether it were possible to determine if skin, which had for many years been exposed to the air, were human or not? I replied, I thought it would be possible if any hairs were present. He then spoke of a friend of his, Mr. Albert Way, whose name is so distinguished in the antiquarian world as one of our first archaeologists, who was desirous of knowing whether certain specimens of skin, stated to have been taken from persons who had committed sacrilege, and which for centuries had been attached to the doors of churches, were unequivocally human. In reply to this, I stated again that if hairs were present, I had no doubt but the discrimination would be comparatively easy.

The following are excerpts from the letters.

*Letter no. 1, May 24, 1847*

> Sir,
>
> ... A tradition exists in Worcester that a man having been caught in the act of committing robbery in the cathedral was flayed, and his skin nailed upon the doors as a terror to the sacrilegious. The doors have been recently replaced by new ones, but they are still to be seen, and having written to a correspondent at Worcester to ascertain whether this strange tale were still remembered, he has sent me a portion of the skin, which is now only to be found under the iron hinges and clamps of the door. One small portion is enclosed, the inner side of which appears to have received the impression of the grain of the wood of which the door was formed. Would you have the kindness to tell me whether you can form an opinion from such a fragment as to the probability that it be human skin or not? The inquiry may perhaps appear trivial, but as a similar tradition is to be found in two other places in England, and no reasonable cause can be suggested why the door of a church should be covered with skin, except from such motive as has been assigned.
>
> I remain, Sir, yours faithfully,
> Albert Way

> On this specimen ... I succeeded in finding two hairs, and thus communicated to Mr. Way the result of their examination, 'I have carefully investigated the portion of skin which you forwarded to me for my inspection, and beg to inform you that I am perfectly satisfied that it is human skin, taken from some part of the body of a light-haired person, where little hair grows. A section of the specimen, when examined with a power of a hundred diameters, shows readily that it is skin, and two hairs which grow on it I find to be human hairs, and to present the characters that hairs of light-haired people do.' [JTQ]

*Letter no. 2, June, 4, 1847*

Dear Sir,

You have now excited my anxiety greatly to obtain specimens of the skin in the two places to which I alluded [Hadstock and Copford] ... I can only state that Sir Harry Englefield accredited and communicated the tale to the Society of Antiquaries, a good many years since, and that when I was last in Essex, it was not forgotten. As regards Worcester ... there is another fragment of this skin in the collection formed by the late Dr. Prattinton, of Bewdley, bequeathed by him to the Society of Antiquaries, and this first drew my attention to it.

I remain, dear Sir, yours very truly,
Albert Way

*Letter no. 3, July 14, 1847*

Dear Sir,

I am enabled to forward to you a fragment ... of the skin, traditionally supposed to have been that of a Dane, attached time out of mind to a door of the church of Hadstock, Essex, supposed to have been pillaged by him. You will very much oblige me if you can, according to your kind promise, aid me in verifying this singular tale, as you did so successfully in regard to the story at Worcester ...

I remain, dear Sir, yours truly,
Albert Way

On this specimen I found three hairs without much difficulty, and wrote to Mr. Way as follows:- 'I have been again fortunate in making out the specimen of skin you last sent me to be human; I found on it three hairs, which I have preserved. I should further state, that the skin was in all probability removed from the back of the Dane, and that he was a fair-haired person.' [JTQ]

*Letter no. 4, Aug. 31, 1847*

My dear Sir,

I have been fortunate enough to obtain from the Incumbent of Copford, Essex, a portion of skin taken from the church door, and to which the like strange tradition had been assigned as in the two former cases ... Will you permit me to state them on the authority of the microscopic examination which you have had the kindness to give to these specimens? As soon as you may favour me with a verdict upon that now submitted for your kind consideration, I shall prepare a short statement for publication in our Quarterly Journal. If the portions of skin to which these strange traditions, long preserved, have by your friendly aid, been authenticated, should appear worthy to be deposited in the valuable collection under your charge at the College of Surgeons, I shall be much gratified in placing them at your disposal.

I remain, dear Sir, yours very truly,
Albert Way

On this portion of skin I found numerous hairs, which were both larger and darker than those from the other fragments, and to Mr. Way's communication I replied as follows: 'I am happy to tell you, that I have succeeded in making out the Copford specimen to be human, as well as the others: I have shown the hairs from this ... to some friends who were sceptical, but they are now quite of my opinion.' [JTQ]

*Letter no. 5, Sept. 24, 1847*

My dear Sir,

... Mr. Neville writes to me, that Mr. Towneley, the rector of Hadstock, had just obtained a scientific opinion in regard to the skin from that church, fully in accordance with your decision. I have enclosed the specimen from Hadstock – as you have been pleased to regard these relics as of sufficient interest to merit a place in the precious collection under your care ... Regretting that they are so trifling in dimensions, I am gratified to be able to add to my sincere thanks any token both of my esteem for your kindness, and of my satisfaction at this alliance and cooperation betwixt Science and Archaeology.

I remain, dear Sir, yours very truly,
Albert Way

For the information of those who may be called upon to undertake similar investigations, I will here state the mode of manipulating which was adopted in order to obtain the hairs from the three specimens of skin above described. The upper surface of each specimen was carefully examined as an opaque object, with a magnifying power of forty diameters, and as soon as a hair was found it was seized with a pair of very fine-pointed forceps and torn out; the hair was then placed between glass, either with or without fluid, and viewed with a power of two hundred linear, which will be found quite sufficient to exhibit all the characters so fully known to microscopists (Fig. 15).

The specimen from Worcester, on its under surface, shows the markings of the grains of the wood, and the paint with which the door was covered; this would go far to prove that the skin was laid on when in a moist state, or soon after its removal from the body: but neither of the other specimens exhibits the same appearances. [JTQ]

Quekett published a note in 1849 on the sample of leathery material found under a metal plate on a church door at East Thurrock, which he identified as the skin of a light-haired man, seemingly confirming a local story of a Danish raider caught in the act and flayed alive, his skin being nailed to the door under the plate.[22]

## False affirmation of the 'Dane-skin' legend

Now that scientific 'proof' of excoriation had seemingly been obtained, its acceptance amongst antiquaries was widespread, and four scholars felt encouraged to publish papers on flaying in medieval England. The first was by Albert Way in 1848,[23] the second by George Tyack in 1898,[24] the third by Harold St George Gray in 1906[25] and the last by Michael Swanton in 1976.[26] Tyack's paper was more romantic than scholarly, and steeped in folklore. The other three authors all cited recorded or alleged acts of barbarism in classical and later antiquity that involved scalping, scourging, flaying and disembowelling. However, it is not always clear which of these acts was performed. Also severe loss of skin could occur through repetitive scourging, which is not the same as flaying a living person, or the de-skinning of a corpse. Very few instances of potentially verifiable flaying are recorded in Britain, and they relate to specific acts of vengeance carried out, or ordered, by a particular person. They were never part of the regular judicial process.

Swanton devoted the first half of his paper to the subject of scalping and bodily excoriation in the western world, citing examples from the random atrocities of barbaric individuals and groups. In some communities it appears that flaying alive was considered a just punishment for anybody who killed his lord: as Swanton put it, 'On the Continent there is ample evidence that the practice of excoriation for *laesa majestas* was no mere fictional horror at the time'[27] (referring to the 12th and 13th centuries). However, neither he nor any other commentator was able to cite an instance of excoriation as a form of English judicial punishment. There is thus no verifiable historical context in England – in civil or ecclesiastical law – for flaying or for the display of human skins on church doors.

Quekett's pronouncements needed re-examination. Swanton wisely opined that a re-assessment of the material evidence was essential, in the light of a century of further scientific development. But the evidence-base had become depleted as a result of the total loss of some samples. Meanwhile, other scholars continued to support, albeit cautiously, the notion that *some* skins on church doors were human.

In 1959, Dr Michael Ryder of the Animal Breeding Research Organisation at Roslin (E. Lothian) examined a fragment of hide reputedly from the Westminster vestibule door that was then in an undisclosed private collection. It displayed characteristics incompatible with a human origin and Ryder concluded that it was probably calf skin.[28] In 1967, Dr Arnold Taylor took up the case of the Westminster skin.[29] He initiated a correspondence with Dr Ronald Reed of Leeds University, Department of Food and Leather Science, explaining the legendary background to the Westminster door, and his wish to obtain definitive evidence with regard to the origin of its hide covering.[30]

Taylor also made enquiries about skin on Scandinavian doors, paid a visit to Sweden and corresponded with Aron Andersson at the Swedish Museum of National Antiquities, Stockholm. The latter reported that no scientific research had, to the best of his knowledge, been conducted on the subject of leather-covered medieval doors, and it was generally believed that only one example survived in the country, from Högby church, Östergötland (Fig. 136).[31] Andersson sent a small sample of the covering from that door.[32]

Taylor submitted this and three samples from Westminster to Reed, whose report ambivalently concluded: 'the indications are that the fragments are human, or from a large animal like a cow or horse or donkey. But the rough grain pattern more favours a human skin'.[33] He added a further relevant comment: 'if you can *definitely* say that the large door, 6 ft × 3 ft, was originally covered with *one*, continuous hide, then clearly we can rule out a human origin'. Reed also commented on the Swedish sample, which he described as 'uncompromising material ... It is vegetable-tanned but it has been so highly polished and compressed that its upper (grain) surface is no longer present; hence one cannot identify the animal from which it came. I also cut transverse sections; again these showed a highly compressed structure with no trace of how the hairs were originally inserted into the dermis. The depth of the papillary layer ... would indicate a heavy calf skin or a light cattle hide'.[34]

Andersson offered his view on the subject:

> To my mind the use of human hides on medieval church doors would be out of the question in any European country. I think the use of human hides for decorative purposes is a novelty in Europe, characteristic of our own civilized century.[35]

Taylor responded:

> Of course I am in entire agreement with you. What I am trying to do is to obtain scientific evidence, on the basic of which one will be able to 'nail' once and for all this absurd legend about the Westminster door. It is most disconcerting when the specialist insists on saying 'he cannot entirely rule out the possibility' that the skin, both at Westminster and Högby, might be of human origin!

Reed's final thoughts on the material from the vestibule door: 'Now, I would guess that the material is cow hide.'[36] Swanton later commented on the unsatisfactory outcome, rejected the skin as being human and identified it as dark, vegetable-tanned cow-hide.[37] In his volume on *Ancient Skins, Parchments and Leathers* (1972), Reed had little to say about the subject of hides on doors:

> An area of interest for the leather scientist concerns the identification of the animal species used to produce a particular material or object. Insignificant as this knowledge might appear, it can sometimes prove of value in many strange directions. As an example one may cite the commonly-held view that in English cathedrals it had long been the practice to face the wooden entrance doors with the hides of notorious criminals. Antiquaries are still deeply interested in this misuse of human skin and indeed there is a fair body of literature on this gruesome topic.

Reed then cited Prattinton's observation about the skin on the north doors of Worcester Cathedral (p. 3), Pepys' diary entry mentioning the west doors at Rochester Cathedral, and the dual traditions relating to the origin of the skin on the vestibule door at Westminster Abbey.[38] He concluded:

> Although the covering of doors with leather was a common mediaeval practice throughout Europe, modern examination of the facings from English cathedrals and abbeys suggests that they are nothing more than cattle hides. Hence no support can

be given to the popular tradition that the hides of notorious criminals were used in this macabre fashion.

He acknowledged that in the 18th and 19th centuries, when the corpses of executed criminals were handed over for medical dissection, they were often excoriated and the skins put to various uses, including book-binding.[39] But that has no relevance to the study of medieval church doors.

Reed also discussed the variation in the dimensions and thickness of skins, in relation to the size and age of the animal:

> The object should be examined closely to judge the area of the original skin, since large dimensions would indicate a large animal and hence a hide rather than a skin. For example, if an object such as a shield or a door-covering had a linear dimension greater than about 3½ to 4 ft, one would have to think more in terms of a fairly mature animal with hide qualities, rather than a skin from a younger animal.
>
> In this connection, one may again refer to the material covering many English cathedral doors and commonly regarded as from the hides of human malefactors. From a man of average build and about six feet tall, pieces that are larger than 3 ft by 3 ft, when processed, tanned and trimmed, would be difficult to obtain. Hence if the door-covering contains pieces greater than about 3 ft along its edges, it is unlikely to have come from a human animal. In many cases of English cathedral doors this indication alone that older (veal) calf skins or more mature cattle hides, from which larger areas of leather would be possible, were indeed used for this purpose, and that the use of human skin is highly unlikely.[40]

Reed made no comment about the Hadstock and Copford skins, presumably because he had not seen them at the time of writing his book. Swanton recorded that fragments of the Copford and Hadstock skins were examined by Reed in 1973–74, which he considered 'likely to be human in origin'.[41] This represented a major change of direction from 1972, when Reed was clearly of the opinion that church doors were covered only with animal hide. In 1973 he examined one of the pieces of Copford skin held by Colchester Museum (Fig. 105).[42] He described it as relatively thick, with dark hair, and was untanned. Hence it was not leather, but a parchment made by stretching the wet, dehaired skin during its drying. The hair-side had been placed against the timber of the door and the flesh-side coated with a gesso-like substance, prior to being painted red.[43] This is erroneous: the paint is on the smooth outer face of the hide. Taunton Museum (Som.) possesses a single, triangular piece of rough skin that was allegedly from Copford, and was donated by Quekett.[44]

Reed examined the Hadstock skin in 1974 and reported on it for Geddes.

> Under the low-power microscope original hair, still in position, could be seen; this allowed the grain and the flesh surfaces of the original skin to be recognised. The grain surface (*i.e.* that containing the hairs) was coated with a mastic-type sealing material apparently used to achieve good contact with the underside face of the hinge. The sealing material is probably a plant-based resin; it is dark brown in colour but contains bright-red granules.

The contact between wooden door and the hinge was extremely good and water-tight since the rust-stained regions around the bolt holes were not extensive. Also, the small areas of the grain surface not covered by the sealing material were very clean and white in colour, indicating a parchment rather than leather.

The flesh side of the original skin appears to have been in direct contact with the wood of the door. It is very smooth and contains only traces of (dried) fleshy meat. Its general cleanliness and light colour confirm that the skin is parchment, for no brown colouration due to vegetable tannin is present.[45]

Here, the remaining hairs were white or light yellow, and Reed stated that they were consistent with being human, observing that 'cattle, sheep, goat and pig are of a different character'. He further observed that the door was 'protected from rain, etc., since the skin sample still retains the pronounced flattened layer structure characteristic of parchment. If the material had received much wetting, this flattened structure would have been destroyed'. A deep, masonry porch has protected the doorway from rain, since the 15th century, but it may have had a timber predecessor.

Swanton was fully convinced by Reed's report that the Copford and Hadstock door skins were indeed human, concluding his paper about 'Dane-skins' and the practice of excoriation in medieval England:

The whole question must therefore be re-opened. It certainly cannot be dismissed, as one might have wished, as a persistent folk-myth. There is now clearly sufficient modern scientific evidence to support the general tradition of excoriation in medieval England, however, much this may have been ignored, or even shunned, in the records.[46]

## Laying the 'Dane-skin' myth to rest

Geddes summarized the occurrences of skins and quoted the opinions of the authorities who had been consulted, principally Quekett, Ryder and Reed. Not unreasonably, she accepted their conclusions, closing her discussion of the subject: 'The occasional use of human skin adds a gory overtone to the mild instructions given by Theophilus.'[47]

Although widespread acceptance of the distasteful fact that at least some of the skins used to cover medieval doors were of human origin has continued down to the present day, there has been growing scepticism by scholars since *c.* 1970. We have quoted the robust views expressed by Taylor and Andersson; other writers also doubted the human attribution, favouring instead cow hide.

In 2001, the Hadstock skin was subjected to DNA testing and the result was scientifically definitive: it is bovine. Twenty years later, another advance in genetic studies, focusing on the peptide content of collagen samples, was developed in the Department of Genetics at Cambridge and has now been applied, non-destructively, to the hides recovered from church doors, and the 'Dane-skin' myth has finally been laid to rest.

## Identifying the species of the hides

*by Ruairidh Macleod*[48]

The macabre nature of 'Dane-skins' – that they comprise the flayed skins of slain Danes or other Vikings, or miscreants – has drawn considerable scientific attention over the past centuries to establish the veracity of their grisly purported origin. Accordingly, this scientific interest has relied on a range of approaches through these periods, beginning with the descriptive recording of the 'Dane-skins' and their associated folklore by antiquarians, followed by the application of microscopy in the 19th century, then detailed comparative morphology of hair and skin structures by the 20th century, and finally the use of biomolecular approaches to analyze DNA and proteins from the 'Dane-skins' at the start of the 21st century. These have invariably centred on the putative identification of the species of origin of the skins, with the aim of thereby proving or disproving the legend in each case, that the remains of the skins are human, and thus a gruesome relic of the punishments that could be inflicted on those who transgressed against the fabric of the church or community in early medieval England.

My own interest in 'Dane-skins' comes from having undertaken a survey of the extant remains of these skins in museums and churches, together with a number of colleagues, over the past four years.[49] The full scientific results of this will be published at a later date, alongside all raw datasets generated in the process, to include findings from the 'Dane-skins' of Worcester Cathedral and Pembridge church, which are currently the subject of ongoing analysis. Here, I discuss the preliminary results of this work to identify the species origin of 'Dane-skins', in the context of the long and storied history of the scientific efforts to corroborate the various sources that describe their human origin. In particular, the application of biomolecular approaches appears now to indicate that their macabre origins may have been somewhat exaggerated.

The first systematic survey undertaken by John Quekett, as mentioned above, found emphatically that the three sets of 'Dane-skin' samples from Hadstock, Copford and Worcester Cathedral did indeed come from light-haired humans, matching the profile of Danish marauders, which in each case had been supplied by local legends. By the 20th century, however, scientific opinion had begun to turn. In 1959, Dr Michael Lawson Ryder, a research specialist in wool, undertook an examination of a piece of the 'Dane-skin' from the vestibule door of Westminster Abbey, and concluded that as the follicles lacked pigment entirely, and were arranged much more densely than would be expected for a human scalp, the sample was most likely from calf skin.[50] Similarly, in 1970, Dr Ron Reed (a highly noted authority on leather and parchment at Leeds University) concluded that another sample from Westminster Abbey was vegetable-tanned cowhide.[51] Subsequently, however, Reed appears to have analyzed samples of the Copford and Hadstock 'Dane-skins' and concluded that these were indeed likely to be of human origin. In a report held by the Saffron Walden Museum, sent by Reed, he noted that the grain pattern and hair distribution of the Hadstock

'Dane-skin' corresponds much more closely to human skin than to either cattle, sheep, goat or pig, and that while the hairs are mostly not pigmented, a few yellow ones were observed. He concluded that the sample is from a parchment preparation of probably human skin from a person with fair or greying hair.[52] By the end of the 20th century, therefore, it seemed that there may be some truth to the origin of the 'Dane-skins' for at least the Copford and Hadstock samples, though with some reasonable equivocation for a method based on comparative scrutiny of morphological characteristics (and thereby not so dissimilar from the approach employed by Quekett).

In 2001, twenty-seven years later, the Hadstock 'Dane-skin' would be analyzed again, through an entirely different approach – by sampling for ancient DNA potentially preserved within the skin. At the time, Dr Alan Cooper (a specialist in DNA, then at Oxford University) was collaborating with BBC documentary makers for the series *Blood of the Vikings* which sought to elucidate the genetic legacy of Vikings upon the UK. For this, a 1.5 × 0.5 cm sample was destructively analyzed by Tom Gilbert, then the student of Cooper. He undertook a Polymerase Chain Reaction (PCR) amplification designed specifically to identify DNA from the mitochondrial control region, a non-coding section of the human mitochondrial genome; this failed to show any DNA from either the sample or the negative controls. Conversely, PCR amplification targeting the equivalent DNA sequence from the bovine mitochondrial genome did yield a positive result, leading to the conclusion that the sample was from a cow. Considering these results now, the findings can be caveated by the fact that at the time PCR amplification was a technique which was extremely sensitive to laboratory contamination, and cow was a common contaminant in laboratories, with cow products found in many molecular biology lab reagents. Deeper DNA sequencing afforded by next-generation of sequencing technologies (not available at the time) would have afforded a more comprehensive result on the species' DNA identifiable from the sample. Nonetheless, this result likely served at the time to reduce confidence that any of the 'Dane-skins' were human, given the positive identifications by both Reed and Quekett of the Hadstock 'Dane-skin'.

The approach employed by myself and colleagues in re-analyzing the Dane-skins differs from that adopted by Gilbert and Cooper in 2001, in that instead of using the DNA sequence to identify the species, the amino-acid sequence of the peptides in the protein collagen are utilized, in an approach described as peptide mass fingerprinting. Specifically, this utilizes a method called ZooMS (Zooarchaeology by Mass Spectrometry), whereby collagen, the most abundant protein in many animals, is extracted from archaeological or historical animal remains and exposed to an enzymatic digestion using trypsin, which predictably cleaves the peptides (chains of amino acids that make up the collagen) at known sites in the amino acid sequence. These fragmented peptides are then analyzed through MALDI-ToF-MS (Matrix-Assisted Laser Desorption/Ionization Time-of-Flight Mass Spectrometry), whereby the mass-to-charge ratio of each fragmented peptide is measured. The peaks for the mass-to-charge ratios can then be compared with predicted distribution of known animal species' collagen peptides digested in a similar way, which vary by species or genus according to their amino acid sequence, allowing species identification.

Given the nature of the remaining 'Dane-skins' having been heavily sampled for the goal of scientific species identification (as well as for the collection of curios) over the past few centuries, a non-destructive approach to sampling was applied (eZooMS). For this, proteins are extracted from skin, leather or parchment samples by gently rubbing a PVC eraser against the sample's surface to generate an electrostatic charge which binds peptides, without necessitating the destruction of the sample material.[53] This was undertaken for the same piece of the Hadstock 'Dane-skin' now at Saffron Walden Museum sampled by Gilbert and Cooper; for three separate pieces of skin held at Copford Church, and for a small fragment that had remained *in situ* under the ironwork of the vestibule door at Westminster Abbey.

Where possible, multiple sites on the skin surface were tested, and the first superficial set of eraser rubbings was discarded as more likely to contain a greater proportion of contamination. For the Hadstock 'Dane-skin', this confirmed the result that the hide was cow, with a very clear set of mass-to-charge ratio spectra supporting this result. For all of the three pieces of the Copford 'Dane-skin' tested, this showed an equine origin; either horse or donkey, given the specific rare peptide which distinguishes the closely related species was not identified. The Westminster sample produced a more equivocal result, however; the mass-to-charge ratio spectra appeared more like a palimpsest of peaks identifying both horse and cow. Further analysis using shotgun proteomics (through liquid chromatography tandem mass spectrometry) is being undertaken to resolve this, although it seems likely that this sample is also equine, with a contamination source of material from a cow. However, none of these showed any indication of peaks specific to human collagen.

From these results so far, it appears that both the principal candidates that Reed identified as likely giving credence to a retributory origin (the Copford and Hadstock 'Dane-skins') are not, after all, human. This strongly suggests that the myth of the 'Dane-skins' may have little basis in historical reality. Nonetheless, in a very separate published investigation of the veracity of historical claims of human skin use, a positive result was only obtained in the last samples analyzed (using a similar approach to ZooMS above). Forty-five Scythian leather samples were analyzed from eighteen burials in Ukraine[54] and, finally, in the case of the last two samples studied, these were identified as from human skin treated as leather, supporting the claim from Herodotus that Scythians made their archers' quivers from the skin of their slain enemies. However, in this case, it seems considerably less likely, particularly given the complete absence of any comparable historical sources for the origins of the 'Dane-skins'. As mentioned above, this research is ongoing, and full results will be published together with my colleagues.[55]

*Addendum*: the analysis of three skin samples from Worcester Cathedral has returned an equine identification (horse/donkey). May 2025.

# 3

# Westminster Abbey: chapter house vestibule door

Westminster Abbey is furnished with an impressive range of historic doors, and others were lost during 18th- and 19th-century restorations. In 2005, the remaining medieval doors were all examined, with a view to dating them by dendrochronology. One door was of particular interest to the writer, whose attention had been drawn to it in the early 1970s by the carpentry historian Cecil Hewett. At the time, we were both studying the timberwork in St Botolph's Church, Hadstock (Ess.), where local tradition asserted that the building and its north door were pre-Conquest, supposedly dating from *c.* 1020. Hence, this had been claimed as the oldest church door in England; it is discussed in chapter 5.

Hewett mentioned that he had seen a door in Westminster Abbey, which he suspected was older than that at Hadstock because its carpentry differed significantly from all other known Norman doors in England (p. 159). In contrast, the Hadstock door is of a familiar type and the other English examples are considered to be post-Conquest (p. 154). Both doors undoubtedly dated from the 11th century, but whether they were Anglo-Saxon or early Norman was debatable. The Westminster door was effectively unknown to the scholarly community, having never been published or discussed (Figs 10 and 22). One outstanding architectural historian – Stuart Rigold – did, however, notice that the door was of especial interest:

> The door-leaf retains its thirteenth-century hasps and one of two remarkably rough strap-hinges of this period, but it has been cut down from an earlier door, which could well have been part of the original fittings of the dormitory range. It is made up of vertical rebated planks, linked together by tightly fitted trenched batons, top and bottom, and a central baton on the other side, bearing an iron strap with recurved ends over a covering of hide. These features are not inconsistent with an eleventh-century date.[1]

Rigold was the first person to grasp the significance of the door, and made no mention of the human skin legend, probably because he was well aware that medieval doors were sometimes covered with animal hide. In 1978, Hewett included a reconstruction drawing of the door in his paper on 'Anglo-Saxon Carpentry' (Fig. 26),[2] and in 1999 Geddes published it in her corpus of decorative ironwork, cautiously dating it to *c.* 1100.[3] Since the early 2000s, the importance of the Westminster door has become widely known and references to it are appearing in new academic publications.[4]

## Location and setting

Access to the chapter house is located approximately midway down the east cloister walk, where a once-magnificent, 13th-century double portal punctuates the east wall and opens into the outer vestibule (Figs 18 and 19).[5] Beyond that lies the inner vestibule, with an impressive flight of steps up to the main chamber of the chapter house. The outer vestibule is three bays in length and two in width, with a row of Purbeck marble columns on the central axis. From these, and the wall-arches to north and south, spring six bays of quadripartite stone vaulting (Fig. 20). There are stone wall-benches in each bay, except where interrupted by doorways. The ceiling of the vestibule is quite low and carries the floor of the chapter library above; prior to the Dissolution, that was part of the large monastic dormitory.

The easternmost bay of the vestibule has opposing doorways in its north and south walls. Both are primary, the former leading into St Faith's Chapel and the small sacristy (p. 5; Fig. 19, door 3). The principal entrance to the sacristy is, however, in its north wall and interconnects with the south transept. This is where Dart recorded that the entrance to the sacristy was secured by three doors, the middle one of which was covered with hide (p. 6; Fig. 19, door 1). Back in the vestibule, the opposing south doorway is where the second skin-covered door still hangs (Figs 21 and 22). The doorway is low and square-headed, with a shouldered lintel of Purbeck marble and plain chamfered jambs, rebated on the south side to receive door 2. It opens into a small, awkwardly disposed space with a low ceiling, behind the library staircase, and was formerly described as a cellar. It was used as an office by English Heritage in the late 20th century, and is currently a store. The chamber was lit by a tiny window in its east wall, now blocked. In the east cloister walk, immediately south of the entrance to the chapter house vestibule, is a 13th-century doorway, opening on to a flight of steps leading up to the chapter library. This was the site of the day-stair to the monastic dormitory (Figs 9 and 19, door 4).[6]

The chapter house, its inner and outer vestibules, the day-stair and the chamber behind it all date from the mid-1250s and are part of Henry III's monumental reconstruction of the Norman abbey. However, from hereon in a southerly direction the dormitory undercroft is a barrel-vaulted Romanesque structure, dating from the late 11th century, with subsequent alterations. The original undercroft was a major component of the Romanesque east cloister, but one or two bays at its northern end were demolished in *c.* 1250, to facilitate building the vestibule and new day-stair. Consequently, the doorway in the vestibule and the cellar under the day-stair may

*Figure 18: Westminster Abbey, east cloister. Twin portal to the chapter house vestibule. Author, © Dean and Chapter of Westminster*

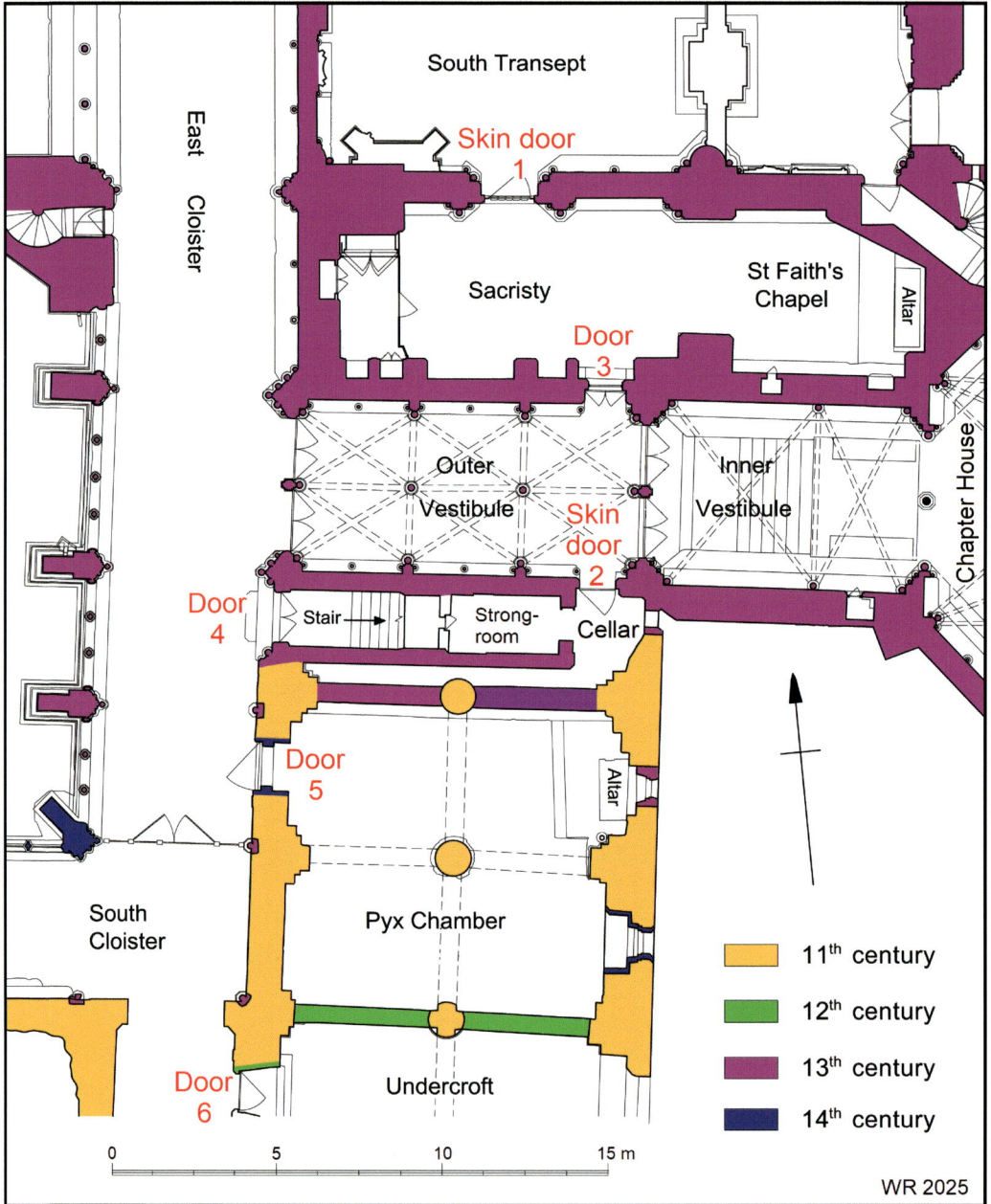

*Figure 19: Westminster Abbey. Plan of the south transept, sacristy, chapter house vestibule and small cellar, showing the locations of important doors (D1–6). Author, © Dean and Chapter of Westminster*

Figure 20: Westminster Abbey. View east from the cloister, into the outer vestibule of the chapter house. The ancient door to the small cellar is on the right, beyond the second wall-pillar. Author, © Dean and Chapter of Westminster

*Figure 21: Westminster Abbey, chapter house vestibule. Mid-11th-century door reused as the entrance to part of the dormitory undercroft (now a cellar) in the 13th century. Malcolm Crowthers, © Dean and Chapter of Westminster*

initially have interconnected with the body of the extensive undercroft. Today, there is a medieval partition wall that completely separates the cellar from the adjacent undercroft, the first two bays of which comprise the Pyx Chamber (Fig. 19). The date of this subdivision is uncertain.

This juxtapositioning gave rise to the Victorian claim that the vestibule doorway was originally the entrance to the Pyx Chamber, but was superseded in the early 14th century, when the present double doors were installed in the east cloister (Figs 9 and 19, door 5). Hence, the 11th-century door was popularly termed the 'Pyx door' (p. 22). Whether this ever served as an entrance to the Pyx Chamber (which became a royal treasury at an uncertain date) depends on when the partition wall of mixed rubble that separates the two spaces was inserted. It is not feasible that access into the north-east corner of the Pyx Chamber, between the mid-13th and mid-14th centuries, was from the vestibule, into the small cellar, and then through a narrow, skewed gap with restricted headroom (owing to the presence of a masonry arch). Chests and other bulky objects could never have been admitted by this route. Also,

*Figure 22: Westminster Abbey, chapter house vestibule. North side of the door to the cellar. This was originally the interior face, reversed in the 13th century. © Dean and Chapter of Westminster*

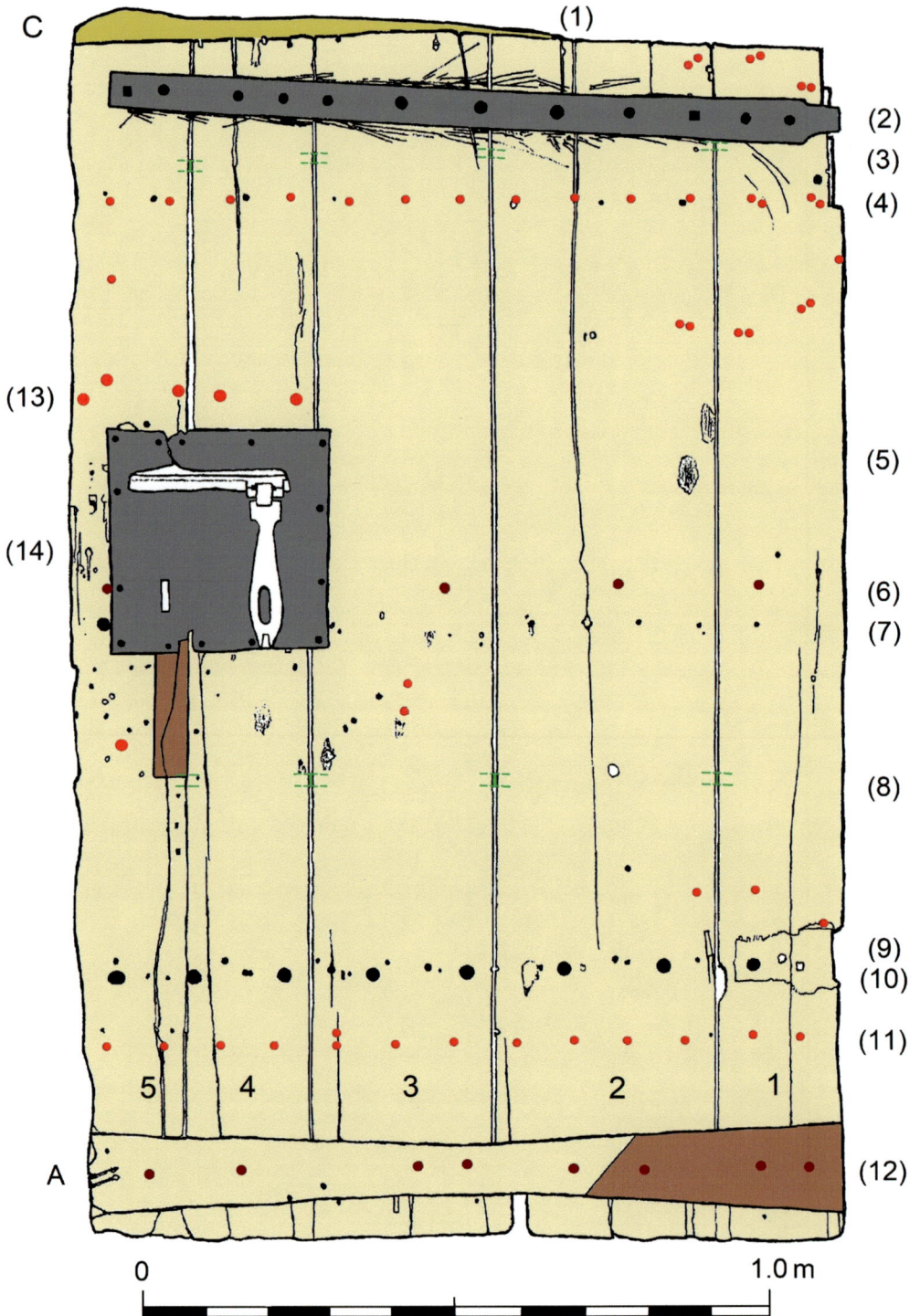

it is surprising that an old, ill-fitting door with low security potential should have been used as the entrance to a royal treasury.

## Form and construction of the door

Since the processes involved in the construction of the door, and the tools required for that task, are described in detail in chapter 4, a brief outline will suffice here. As seen today in its cut-down form, the vestibule door is square topped and measures 1.98 m (6 ft 6 ins) high by 1.26 m (4 ft 2 ins) wide (Fig. 23). It is constructed from five oak boards 40 mm (1½ ins) thick, ranging in width from 22.5 cm (9 ins) to 39.0 cm (15½ ins). The boards are rebated, with a generous lap of 35 mm, and are secured edge-to-edge by two dowels (12 mm diam.) in each joint; these were drilled through the centre-lines of the laps, implying that the holes were made before cutting the rebates. There is likely to have been a third tier of edge-dowels in the lost semicircular head of the door.

The boards were converted from the trunk tangentially, by sawing through-and-through. Slight but unmistakeable traces of the kerf-lines caused by sawing remain on the surface of at least one plank (Fig. 44B). The angle of sawing was 55–60 degrees to the axis of the trunk, which is consistent with the use of a saw-pit.[7] The planks were partly seasoned before use, as evidenced by the small amount of shrinkage between the joints.

The boards were assembled and locked together by their rebates and edge-dowelling (Fig. 23, nos 3 and 8). They were then strengthened by adding three ledges (A, B and C) in a most unusual manner: these were not surface-fixed, as was the normal Anglo-Saxon practice, but set into shallow housings cut into the front and back of the door. Consequently, both faces were flush. There is a single ledge (B) located on the front of the door at mid-height, secured with six pegs, although another may have been lost from the reduced board 5 (Figs 24, 25, 27 and 38). Only two pegs appear to be primary, and they are both in board 1; the peg has been lost from board 4, but

*Figure 23: (opposite) Westminster Abbey, vestibule door. Archaeological analysis of the north face (original interior). Author, © Dean and Chapter of Westminster*

*Key: 1. Remnant of matrix for ledge C; 2. 13th-century strap-hinge, flanked by cut-marks for removing the hide covering; 3. Upper row of edge-dowels joining the boards (green); 4. Holes for clench-bolts attaching the original upper strap-hinge and C-scroll (red); 5. Two candle burn-marks on board 2; 6. Ends of pegs securing ledge B on the south face (maroon); 7. Points of clenched nails attaching the mid-height iron strap on the south face; 8. Lower row of edge-dowels joining the boards (green); 9. Nail-holes formerly attaching a 13th-century strap-hinge; 10. Bolt-heads attaching the 1860s replacement hinge on the south face (black); 11. Holes for clench-bolts attaching the original lower strap-hinge and C-scroll (red); 12. Ledge A and its fixing pegs (maroon); 13. Clench-bolt holes for the attachment of an original locking device; 14. Late medieval iron lock-plate, sliding bolt and renewed hasp. Modern timber replacements are shaded dark brown.*

*Figure 24: Westminster Abbey, vestibule door. South face (original exterior), seen from within the cellar. © Dean and Chapter of Westminster*

its 15 mm diameter hole remains. The pegs are circular in cross-section, with one exception, in board 1, where it was made from a squared piece of oak, from which the arrises were shaved off, resulting in a sub-rectangular cross-section (Fig. 37). Pegs of this form are typically found in Saxo-Norman joinery, as in the window frames at Hadstock. The right-hand end of the ledge was truncated when the door was reduced in width, and some of the timber was replaced more recently (Fig. 25).

The back of the door was fitted with two ledges, but the top one (C) was lost when the door was truncated in the 13th century, and only the lower edge of its matrix survives (Figs 23, 29A and 58). This ledge was originally positioned on the diameter of the semicircular head (Fig. 26). The third ledge (A) is close to the bottom of the door, and the right-hand end has been renewed (Fig. 28B). It is currently secured with seven pegs, all but one of which seem to have been replaced.[8]

The ledges were thin, rectangular pieces of board, 15 mm thick, with their long sides slightly concave, so that in elevation they were dovetailed at both ends (Figs 23, 25 and 65). The purpose of this was to restrain the boards from spreading laterally. The two surviving ledges were of similar dimensions, but the arcature of their long sides is not identical. When each ledge was fitted in its matrix, flush with the surface, holes were drilled through it and the boards to receive pegs that were driven in tightly. In his first exploded diagram of the door, Hewett showed the pegs as being fox-wedged (*i.e.* the outer end of each peg was given a diametrical saw-cut, and a tiny wedge driven in to tighten its grip). In reality, the pegs are not fox-wedged and Hewett corrected the error in a later diagram (Fig. 26).[9] All but one peg-hole was drilled through the full thickness of the door.

With the carpentry complete, it remained for the faces of the door to be planed or trimmed with a shave, to obtain a perfectly flat finish, free from minor steps and irregularities where the boards abutted one another. It also facilitated the removal of rough surfaces left by sawing the planks; this was specified by Theophilus in his early 12th-century treatise on door construction (p. 148; Fig. 44A). Traces of the finishing process have been recorded (p. 74). The door was now ready to receive its covering of animal hide.

We see the door today in a cut-down form. The width has been reduced by at least 11 cm, a little taken from the hinge-edge (board 1), but more from the closing edge (board 5). The door has also been considerably truncated in height, leaving only residual traces of the matrix that once held the uppermost ledge (Figs 29A and 58). If the door in its original form had been square-topped, it must have been about 15 cm taller than it now is. However, this was clearly an important door and it is more likely to have had a semicircular head, as reconstructed by Hewett and endorsed by other commentators.[10] Its original height may therefore be estimated as *c.* 2.75 m (9 ft), and the width as 1.37 m (4 ft 6 ins) (Fig. 36B). The hinges, albeit now represented only by scars, confirm the latter dimension as the minimum width of the door. These measurements define a door with the common height-to-width ratio of 1:2.

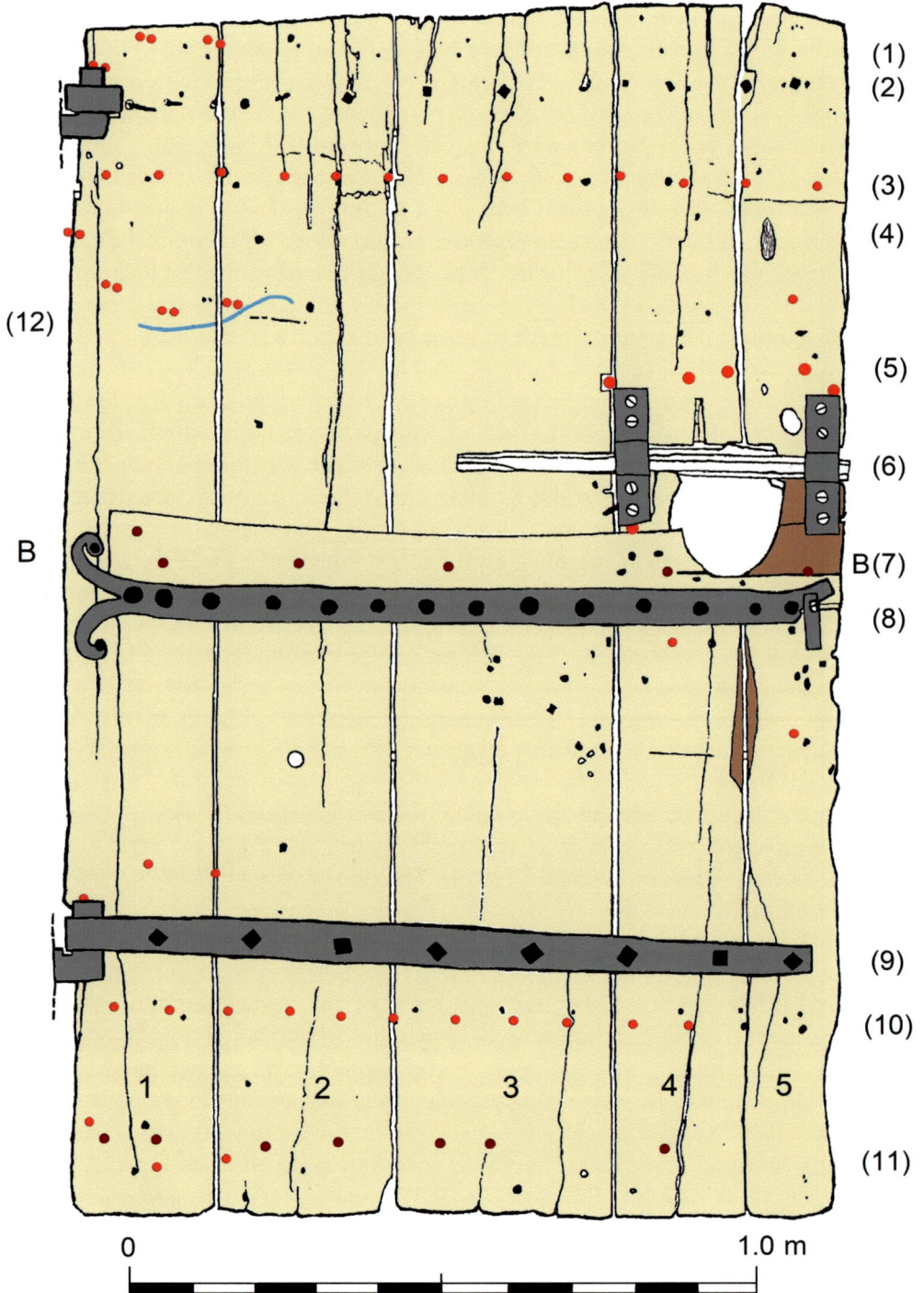

(1)
(2)

(3)

(4)

(12)

(5)

(6)

B                                                                    B(7)

(8)

(9)

(10)

1          2          3          4          5

(11)

0                                                                1.0 m

*Figure 25: (opposite) Westminster Abbey, vestibule door. Archaeological analysis of the south face (original exterior). Author, © Dean and Chapter of Westminster*

*Key: 1. Line of small tack-holes, formerly securing the junction between two pieces of hide; 2. Fixings for the 13th-century strap-hinge on the north face, comprising original clenched nails and modern bolts; 3. Holes for clench-bolts securing the original upper strap-hinge and C-scroll (red); 4. Apotropaic candle burn-mark on board 5; 5. Clench-bolt holes for original locking device (red); 6. Late medieval sliding bolt with replaced bearing-blocks; 7. Ledge B and its pegs (maroon); 8. Original mid-height iron strap and fixing nails; 9. 1860s replaced lower strap-hinge and pintle; 10. Holes for clench-bolts securing the original lower strap-hinge and C-scroll (red); 11. Ends of pegs securing ledge A on the north face (maroon); 12. Scored line associated with marking the position for the C-scroll of the upper strap-hinge (blue). Modern timber replacements are shaded dark brown.*

## The hide covering

The use of animal hides for covering timber artefacts, such as doors, trunks and shields, has a long history. The Laws of Æthelstan (AD 926–30) stipulated that cow hide was to be used for covering shields, rather than sheep skin.[11] The form of construction employed at Westminster, which resulted in the door being flush and smooth on both faces, is without parallel in medieval Britain. Normally, hide would have been applied only to the exterior face, but this being a double-sided door, facilitated treating both faces the same. Confirmation that this was so was first provided by Scott – a reliable and astute observer – when he reported in 1859 that the door retained evidence of having been covered with skin, 'within and without' (p. 12).

Over the course of time, the covering became tattered and was entirely removed from both faces of the door. A sharp knife was drawn along the edges of the ironwork, to cut through the hide, and the marks left by these excisions are visible on the faces of the boards (Fig. 29B). On the north face,

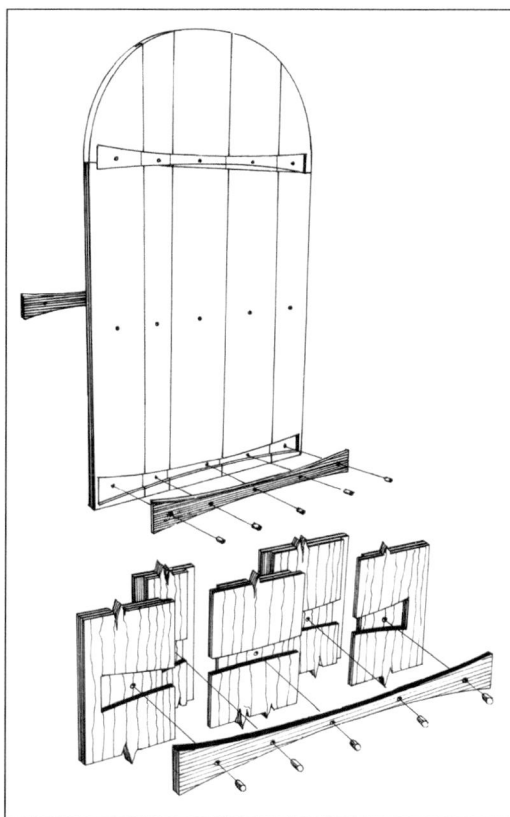

*Figure 26: Westminster Abbey, vestibule door. North face. Exploded diagram to illustrate the components of the door. Hewett 1985, fig. 149*

*Figure 27: Westminster Abbey, vestibule door. South face. Central area of the door, showing inlaid timber ledge (B), wrought iron strap and security bolt. The pale areas indicate where the boards were scraped in the 19th century, probably to remove the residue of glue that had originally attached the hide covering. Author, © Dean and Chapter of Westminster*

however, along both edges of the upper hinge are multiple slashes marks, indicative of a more frenzied attack on the hide (Figs 23 and 31).

In 1860, Scott could only report that two tabs of hide hung down from the iron strap on the south face of the door (p. 12). But even after the remainder of the skin coverings had been comprehensively torn off, three strips still survived – one on the front and two on the back – all inaccessibly trapped beneath iron straps. That on the front still lies under the original mid-height strap with split-curl ends, and the cut-edges of the hide are exposed in the narrow joint between the boards and the ironwork (Fig. 30). The two replacement strap-hinges fitted to the north face in the 13th century, when the door was reversed and rehung, also concealed and protected narrow strips of hide, and the upper hinge still does.[12] However, when Scott replaced the lower hinge, the strip behind it was salvaged, rolled up and kept in a small box in the Pyx Chamber. There it remained until the 1980s or 1990s, sometimes being brought out and shown to privileged visitors as an item of antiquarian interest.[13] The whereabouts of this important find has not been traced, and it is presumed lost.[14]

Scientific examination in the 1970s of a sample of the skin concluded that tanned cow hide was the covering material, and in 2022 that was confirmed by collagen analysis (p. 34).[15] The method of attaching the hide to the door is not now apparent, but it was presumably glued in the manner prescribed by Theophilus (p. 148). On the outer face of the door, the hide was additionally secured by the ferramenta that were next attached. Since no original metal attachments are known to have existed on the rear side, an adhesive must have been employed. The possibility that some tacks were nailed around the vulnerable edges cannot be ruled out, but physical damage and reduction in the overall dimensions of the door have removed any evidence that might once have existed (*cf.* evidence at Hadstock for a fixing-band around the perimeter of the door, p. 102).

*Figure 28: Westminster Abbey, vestibule door. North face, inlaid ledge, close to the bottom of the door. A, left-hand end in boards 3–5, showing its original dovetail-shape. B, right-hand end in boards 1–3, with the modern repair; pegs ringed in white. Both views include (at the top) the lines of drilled holes for attaching the primary lower strap-hinge with clench-bolts. Author, © Dean and Chapter of Westminster*

Several antiquarian writers questioned how many hides were needed to cover one face of the door. The skin of a calf would be far too small, and that of a mature cow would also be insufficient. While there would be no problem with a single cow hide spanning the width of the door, one-and-a-half hides would have been required to cover the full height. If the hides were attached with glue alone, there would be no physical evidence to find, but if the joints were also nailed there should be visible indications on the boards. Scrutinizing the elevations for evidence of horizontal joints yielded compelling results for a line of nailing on the former external face (Fig. 25, 1). It is 15 cm above the position of the upper hinge, and almost at the top of the (truncated) door. Thus, it would appear that the external face of the door was covered with a single sheet of hide (1.95 × 1.37 m), augmented by a semicircular sheet (1.37 × 0.75 m) filling the arched head.

Less certain are details of the attachment of hide to the back of the door (Fig. 23). Two horizontal lines of nail-holes are discernible, and the lower one is 45 cm above the bottom of the door. It coincides with the location where the 13th-century hinge had formerly been attached, and the holes were most likely associated with it.[16] The second line is fragmentary and occurs at the very top of the door and potentially relates to the attachment of a semicircular piece of hide.

Figure 29: Westminster Abbey, vestibule door. Details of the upper part of the north face. A, remains of the matrix for the lost ledge (C). B, part of the 13th-century strap-hinge flanked by the knife-slashes that cut away the hide covering; below is the line of drilled holes for the clench-bolts attaching the original upper hinge. Note, on the right, two pairs of double-drilled holes for fixing the C-scroll. Author, © Dean and Chapter of Westminster

On both faces of the door are other, scattered holes for small nails of unknown purpose. Since there were no iron fittings on the north, and so little evidence for nailing, it forces the conclusion that the attachment of the hide relied on glue. Were it not for the two strips of hide preserved *in situ* under the 13th-century hinge-straps, one might otherwise have concluded that there never was any hide on the back of the door.

Covering both faces with hide took place in two stages, before and after fixing the ferramenta, respectively. The front was covered first, and then the ironwork nailed or riveted in place. Nails were clenched (*i.e.* the points hammered over and flattened into the boards on the back face of the door), and the ends of any rivets would have been hammer-dressed there too. If hide had already been attached, these processes would have done irreparable damage to the covering. Consequently, by adding the hide on the back, after the ironwork had been attached to the front of the door, ensured that the unsightly scars of fixings were fully concealed.

Confirmation that glue was used to attach the hides may be seen at several locations on the south face. When the door was rehung in the mid-19th century, and generally tidied up, a metal cabinet-scraper was used to remove patches of a substance that was adhering to the timber: this must have been the residue of glue. Consequently, scraping removed not only it, but also the underlying patina of the timber, resulting in the pale areas that we see today (Figs 27, 38 and 39).

Although samples of the hide from the door were variously acquired by interested parties in the 19th century, their fate is unknown. However, three small scraps are preserved in the Abbey Collection (Fig. 32):[17]

1. A triangular fragment, heavily blackened with post-medieval paint. Traces of earlier red paint are also present.
2. The middle-sized piece measures 37 mm in length, and is a trimming from alongside a metal strap. The fragment is curled and partly blackened with the paint applied to the iron strap. No traces of red colour were observed.
3. The largest piece measures 63 mm in length. It had been partly under an iron strap, and is creased along its length. This is the result of a detached flap of hide hanging down over the strap for long enough to adopt creasing that reflects the thickness of the ironwork. The majority of the flap was later cut off, leaving just a few millimetres of hide projecting above the edge of the metal. The remaining, curled fragment of hide was subsequently trimmed off by cutting a clean line alongside the edge of the strap at an unknown date, probably in the first half of the 20th century. The surfaces of the hide are smooth and cream in colour, but dirtied; on the outer face, at one end, are traces of a white coating, overlaid by bright red paint (p. 55).

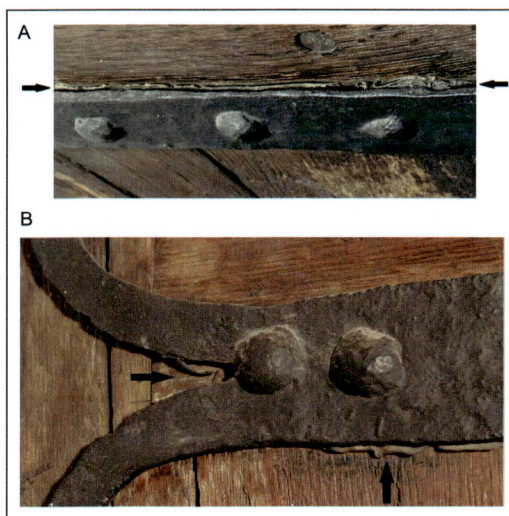

Figure 30: Westminster Abbey, vestibule door. South face. A, oblique view of the top of the mid-height iron strap, showing the cut edge of the hide trapped behind it. B, fragments of curling hide protruding from under the strap (arrowed). Author, © Dean and Chapter of Westminster

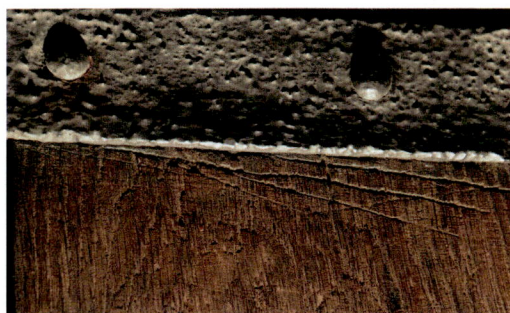

Figure 31: Westminster Abbey, vestibule door. North face. Scoring of the boards alongside the top hinge where a knife was used to trim away the remains of the hide covering. Author, © Dean and Chapter of Westminster

There are also traces of dark red paint, which visually matches that on the stone reveal (west) of the doorway to the cellar. This appears to be part of the primary decoration of the 13th-century vestibule. Several curling slivers of hide, similar to sample 3, still project from beneath the lower edge of the central iron strap (Fig. 30B).

*Figure 32: Westminster Abbey, vestibule door. South face. Fragments of hide recovered from alongside the mid-height iron strap. In the early 20th century these were displayed in the Abbey Museum, as the pin-holes bear witness. A, outer face of the hide, contaminated with black paint from decorating the adjacent iron strap. B, underside of the hide (flesh face). Author, © Dean and Chapter of Westminster*

### Paint layers on the hide
*by Krista Blessley[18]*

The fragments of hide were examined in the Conservation Studio.[19] They had previously been identified as cow hide.

Paint examination of three small fragments of skin has been undertaken at the request of the Abbey's Consultant Archaeologist. Owing to the small size of the surviving paint it is not appropriate and thus not recommended that any destructive analysis techniques be undertaken (*e.g.* paint cross-sections, pigment dispersions or SEM EDX). The

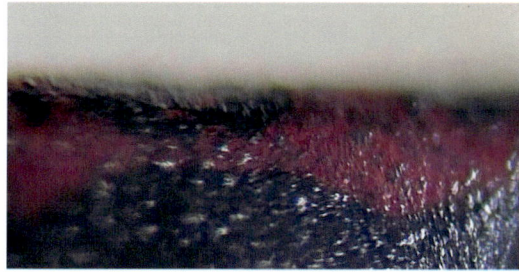

*Figure 33: Westminster Abbey, hide sample 1. Dark red underlayer visible at the edge, beneath the black paint layer. Incident light, SPF 300 F. Krista Blessley, © Dean and Chapter of Westminster*

following is therefore based solely on surface examination using stereomicroscopes (Leica and SPF 300 F) and a digital microscope (Dinolite).

#### Sample 1
This piece of hide exhibits a black paint layer (similar to that found on the others: see below) which extends from the recto slightly beyond the turnover edge (Fig. 33). At the edge of the hide it is possible to see a dark red underlayer, the colour of which suggests a red earth-containing pigment. It appears similar to the dark red paint identified on sample 3.

#### Sample 2
This sample also has a black paint layer. It extends onto the verso of the skin at the edge. It was applied directly onto the hide with no priming or ground layers. It has a patchy saturation and in some areas appears quite matt and underbound. These differences in saturation may relate to a surface coating or variations in the hide.

#### Sample 3
The largest of the three hide samples has the most surviving paint. A white ground layer (possibly chalk and/or lead white) is visible on the verso (Fig. 34). On top of this ground layer are fragments of a red paint layer. This layer appears to be red lead-containing in colour (untested) and painted directly on the white ground (Fig. 35A). There may be a further, red paint layer on top (more clearly visible when the paint layers are wetted out with white spirit), but this is heavily obscured by the surface dirt, dust and accretions, so it is difficult to ascertain with any certainty.

The surface of this red lead (untested) paint layer is covered with protrusions, likely to be lead soaps. These are hybrid compounds, containing both fatty acid chains and metals associated with a carboxylate function, usually thought to result from an uncontrolled reaction of oil with lead-based pigments, in particular lead

*Figure 34: Westminster Abbey, hide sample 3. Photomicrographs of remains of paint at three locations on the fragment. A, white ground layer, red paint layer (possibly red lead) and further uppermost red paint layer (obscured by surface dirt and accretions). B, dark red paint layer with no ground or priming, possibly red earth-containing; black paint layer on top. C, black paint layer. Krista Blessley, © Dean and Chapter of Westminster*

white, red lead, and lead tin yellow.[20] A further black paint layer (similar to those seen on the other samples) is also present here and is on top of the dark red paint layer (Fig. 35B).

In addition to the aforementioned red paint layers, there is a dark red paint layer, which appears similar to that found at the edge of sample 1 (under the black paint) and may contain red earth pigments. This dark red layer is in a different area from the ground/red-lead layers and appears to be directly on the hide, with no ground or priming.

## Iron fittings

### Clenched nails, roves and clench-bolts: clarification of terminology

The literature on historic carpentry – particularly that relating to doors – is replete with references to these iron fixings, but their usage has become confused, and clarification is needed.

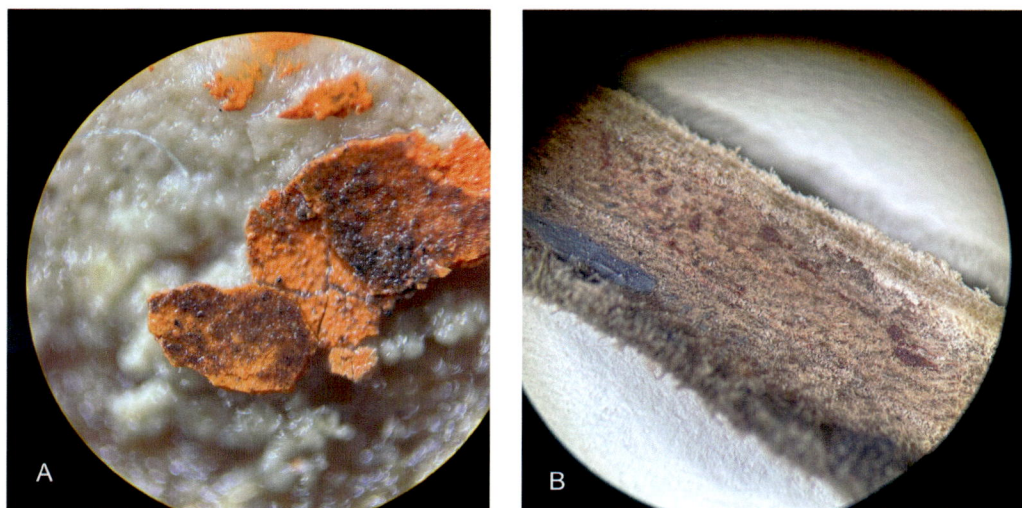

*Figure 35: Westminster Abbey, hide sample 3. A, photomicrograph of the principal area of surviving red paint (for context, see Fig. 34, A). White ground, red lead paint layer (untested); possible further dark red layer, obscured by surface dirt and accretions. B, photomicrograph showing the dark red paint and the overlying black paint (left-hand side). SPF 300 F microscope. Krista Blessley, © Dean and Chapter of Westminster*

Nails may be used to join two pieces of timber together, or to attach a metal fitting. If the nail is longer than the combined thickness of the materials through which it passes, the point will project at the rear. When the point is hammered over, the nail is said to have been 'clenched'. Usually, the point will be flattened into the timber, so that it does not become a snagging hazard (Fig. 41A). The purpose of using nails that are 'too long' is to strengthen the attachment: *i.e.* once clenched, they cannot work loose or be easily withdrawn. However, hand-forged nails can be brittle, and the stress induced by bending the shank can weaken the metal. Over the course of centuries, rust may attack the bend and the point falls off, leaving its indentation in the timber.

The dictionary definition of a 'rove' is simple: 'a metal plate through which a rivet is passed and then clenched over'. In antiquity, the rivet was generally a nail, and roves mostly took the form of square or diamond-shaped washers. Together, the two components formed a 'clench-bolt'. However, in modern usage clench-bolts are commonly, but incorrectly, referred to as 'roves'.

Clench-bolts provided a stronger joint than clenched nails and were a characteristic of clinker-built boats in the Anglo-Saxon, Viking and later periods. A nail with a large head was driven through the overlap between two planks, the point projecting on the inside of the hull. A rove was then slipped over the shank of the nail, which was hammered to spread the projecting metal until it gripped the rove. The boards were thus securely riveted together, and the nail and rove could not be parted. The 7th-century Sutton Hoo burial ship was clinker-built, using many hundreds of clench-bolts.[21]

In relation to the archaeology of doors, the term 'rove' has also become misused to refer to decoratively shaped washers that were slipped on to nails before they were driven into the ledges or framing: the washer simply sat under the head of the nail, and no riveting was involved. This use is exemplified in the north door at Hadstock (Fig. 81).

### Reconstructing the lost ironwork

Returning to the construction of the Westminster door, once the hide had been attached to the outer face, and painted, the ironwork could be added. The vestibule door was fitted with four iron straps, all extending across its full width (Fig. 36B). Two of them were hinges, the others stiffeners. The only primary attachment that survives today is the mid-height strap: it is 55 mm wide by 5 mm thick, with split-curls at both ends, fixed with fourteen large-headed nails, plus two in each of the curls (Fig. 37). The nails were driven through the boards and clenched. Their points have rusted away, but the indentations resulting from the clenching are discernible in the back of the door.

Although the original hinges were removed in the mid-13th century, and replaced by others on the opposite face of the door, the scars where they were attached are readily apparent, as are the numerous fixing holes in the boards (Figs 38 and 39). Again, the hinge-straps were c. 55 mm wide, and both these and their attached C-scrolls were secured with clench-bolts, 6–8 mm in diameter. Each strap had thirteen bolted fixings, and their C-scrolls had eight apiece. The split-curls on the C-scrolls were nailed, as were presumably those at the outer ends of the straps, but the ghost evidence for these was lost when board 5 was reduced in width (Figs 25 and 40).

The loss of a few centimetres from the hinge edge of board 1 has also removed some of the ghosted detail for the curvature of the 'C', but it is clear that the scrolls were more oval than circular in plan, and the lower one was markedly irregular. Both C-scrolls were non-symmetrical in relation to the positioning of their respective hinge-straps, a common feature of hinges where the components were welded together.

A problem clearly arose when fixing the upper hinge. Its 'shadow' on the boards exhibits double-drilling for the eight holes of the C-scroll, but not for the strap (Fig. 39).[22] The paired holes are c. 12 mm apart, indicating the amount by which the hinge had to be repositioned. Logically, the strap would be fixed first, followed by the C-scroll, but that was not the case here. The carpenter started by incorrectly drilling the bolt-holes for the upper C-scroll, his error being spotted before he began to drill those for the strap. The lower hinge did not require repositioning.

C-scrolls, whether welded to hinge-straps, or simply overlaid on them, were decorative features and were normally nailed to the boards in the same manner as the straps themselves. In this instance, however, the fact that the fixings had to be inserted

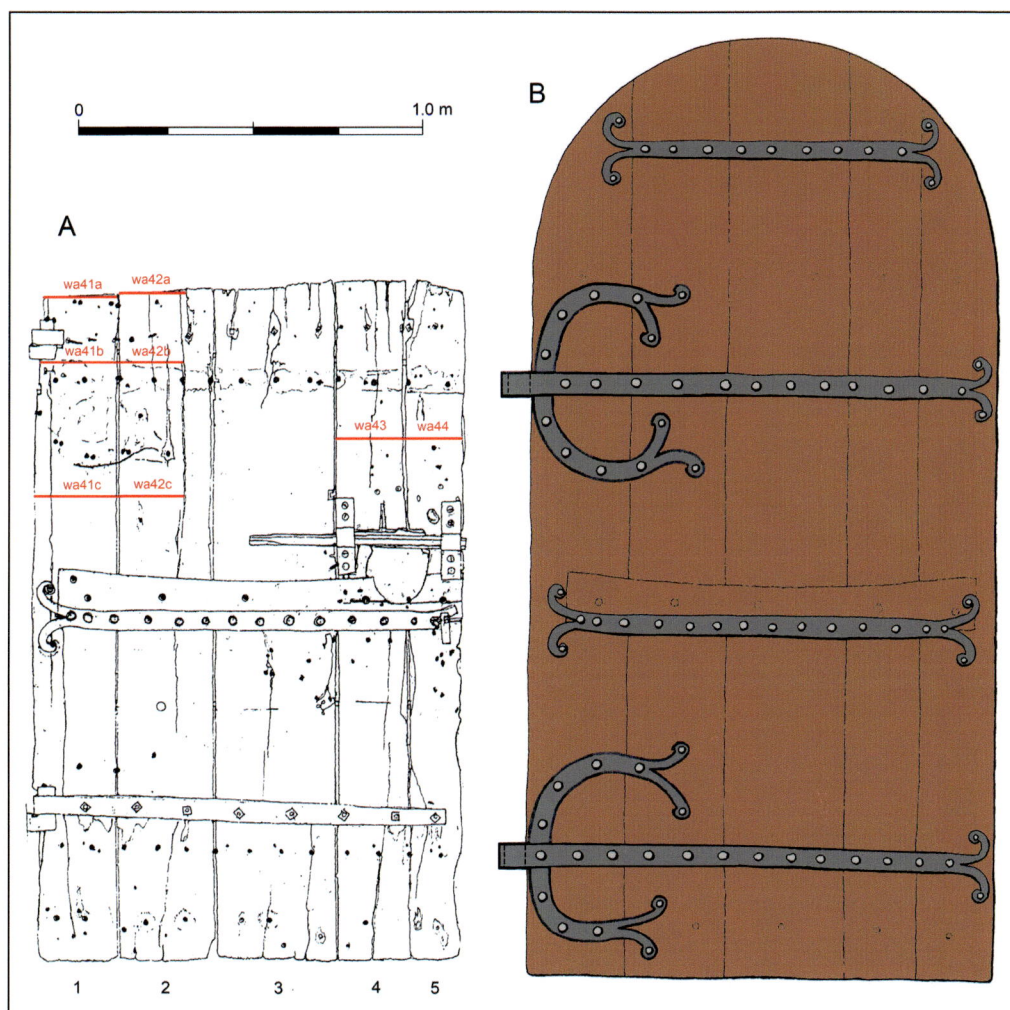

*Figure 36: Westminster Abbey, vestibule door. South (outer) face. A, archaeological record of the elevation, as existing, with boards numbered 1–5; positions of the micro-bores taken for dendrochronology are shown in red. B, indicative reconstruction of the door's features, omitting the hide layer, with the height and width restored to their original proportions. Author and Angela Thomas, © Dean and Chapter of Westminster*

into pre-drilled holes implies that common nails were not used, except for securing the ends of the split-curls. The employment of twenty-one clench-bolts to attach each hinge was excessive, bearing in mind that this was a relatively lightweight door, which did not have a heavy oak frame attached to its rear face. Moreover, a basic clench-bolt did not require a pre-drilled hole when it had only to pass through the thickness of a single board.

*Figure 37: Westminster Abbey, vestibule door. South face. Central iron strap terminating in a split-curl, each end of which was fixed with a single nail; also showing the left-hand end of the middle ledge (B), with two pegs (ringed in white) attaching it to board 1. One peg is circular, the other sub-rectangular in cross-section. © Dean and Chapter of Westminster*

Roves sometimes leave a slight scar on the surface of the timber, but no examples are discernible on the back of the door. Inexplicably, there is one circular indentation, 16 mm in diameter, on the front, which would have been behind the upper hinge-strap, where riveting could not have taken place. The fact that all the holes for the hinge fixings were drilled indicates that conventional clench-bolts were not used. The latter, if inserted through the hinge-straps on the front of the door, and riveted over roves on the back, would have created a series of bumps that disfigured the hide covering. This could be obviated by inserting studs with large flat heads, through the back of the door, and riveting their shanks on the front, where they projected through the hinge-straps. Thus, the hinges were clench-bolted, their straps effectively substituting for individual roves.

The door had a lock attached to the inner face, and evidence for its fixing is partially preserved in the vicinity of the present sliding bolt. Just above this is a row of four drilled holes, and another below the bolt (Figs 25, no. 5 and 38). The door is badly damaged in this area and further fixing-holes have doubtless been lost. The drilled holes are larger (9–10 mm) than those associated with the hinges, and were clearly intended for clench-bolts.

There are sundry other nail-holes and drillings in the door, but nothing to indicate that any primary ferramenta were attached to the back face.[23] The hide there must have been applied after the external ironwork had been fixed to the front, and all projecting nails clenched or riveted, otherwise the interior of the door would have been severely marred by the visibility of nails and hammer-damage to the hide.

*Figure 38: Westminster Abbey, vestibule door. Upper half of the outer (south) face of the door, showing the shadow where the original strap-hinge was fixed. The pale area in the lower part of the C-scroll probably results from glue being scraped off in the 19th century. Author, © Dean and Chapter of Westminster*

## Later interventions with the door

In addition to the physical evidence relating to the construction and initial hanging of the door in Edward the Confessor's abbey, it has suffered mutilation and adaptation over the ensuing centuries. We may reasonably assume that the door hung in its primary location from the 1060s until *c.* 1250, when Henry III instigated the demolition and rebuilding of the south transept, chapter house and the northern half of the Romanesque east cloister range. The chapter house was constructed in the early 1250s, and the windows were ready for glazing in 1253;[24] the great tile pavement would have been laid in *c.* 1255.[25] The inner and outer vestibules must have been under construction at the same time, since they provided the only access to the chapter house. It is likely that work on them continued until *c.* 1260, and hanging doors would have been one of the final operations.

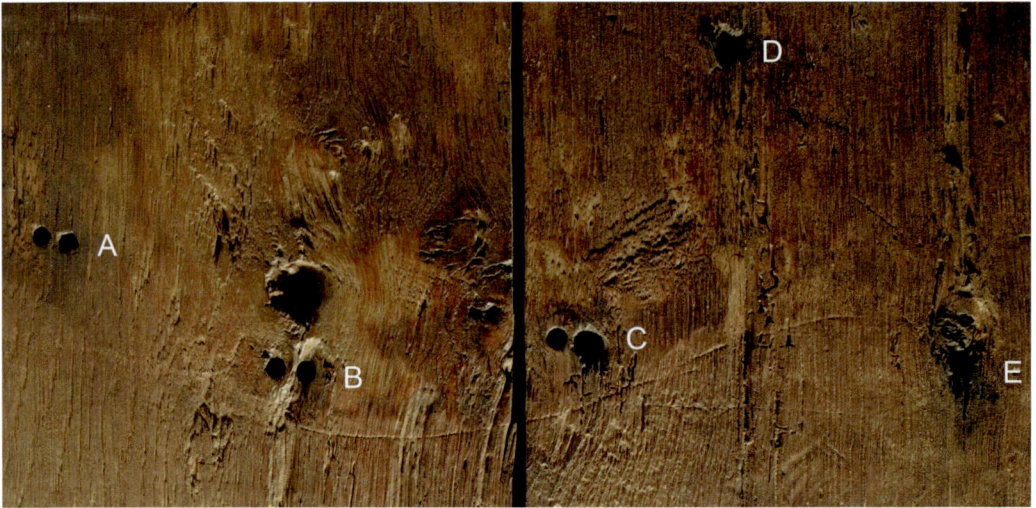

*Figure 39: Westminster Abbey, vestibule door. South face, showing holes for fixing the upper hinge. A, B, C are drilled holes for clench-bolts for the lower half of the C-scroll, and D, E are the remains of nails that attached the terminals of the split-curl. Note also double-drilling of three holes, to correct a placement error, and the single scored line following the curvature of the C-scroll and the split-curl; this was a positioning aid for the hinge before the hide was attached. © Dean and Chapter of Westminster*

The salvaged 11th-century door was evidently to hand and was cut down to the required size for the opening in the south wall of the outer vestibule. It was stripped of its two original hinges, but the iron strap at mid-height was retained; even so, part of the split curl on the outer end of the strap had to be wrenched off in order to reduce the width of the door to meet its new requirement. The external face was now heavily scarred, but the hide-covered back would have been intact and potentially without serious disfigurement. Hence, this was chosen to be the new exterior face, as seen from within the vestibule.

Two plain strap-hinges were fitted to the north face of the door, each being secured with twelve nails driven through the hide and boards, and clenched on the reverse side. The upper hinge is still in place and functioning, but eight of its nails were replaced with bolts and nuts in the 19th century; the strap is slightly tapered along its length, and is cranked so that the hanging point is inside the chamber (Figs 23, no. 2 and 38).[26] The lower hinge was doubtless similar, but has been lost and superseded by a new strap-hinge, which is fixed with eight coach-bolts and square nuts. This is Scott's work, dating from *c.* 1860; both pintles are leaded into the west jamb of the doorway, but only the upper one appears to be original (Fig. 41A).[27] The new hinge was fitted to the south face of the door, and did not have to be cranked like the upper one, giving rise to the curious spectacle of a door with one strap-hinge on each face. Black paint was applied to the ferramenta in the 19th century, and some of this overspilled on to the timber and projecting edges of the hide (p. 53).

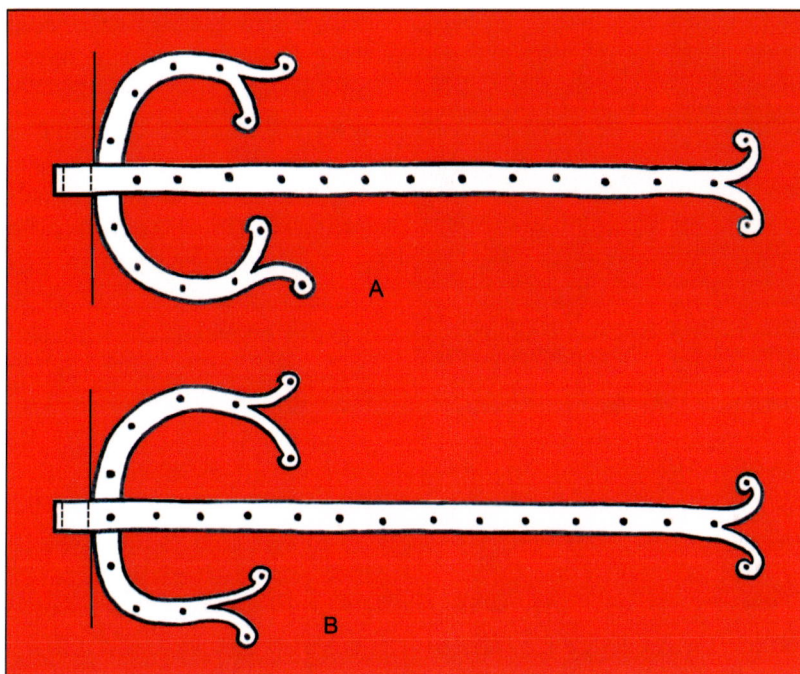

*Figure 40: Westminster Abbey, vestibule door. Reconstructions of the two lost C-scroll hinges, based on shadow outlines and fixing-holes in the boards. A, upper hinge. B, lower hinge. Author, © Dean and Chapter of Westminster*

At mid-height, boards 4 and 5 – and to a lesser extent board 3 – display a multiplicity of fixing scars relating to a complex history of latching and locking devices. Holes for primary clench-bolts have been described above. The present closing device comprises a 35 cm square iron plate with a horizontal slot close to the top, behind which is a sliding bolt that engages with a hole in the east jamb of the stone door reveal. The bolt has a hasp attached, which can be secured to a staple with a padlock (Fig. 42). This locking device is late medieval, but the hasp, staple and bearings for the bolt have all been renewed. The outer plate is now fixed with woodscrews, replacing nails.[28] The plate also covers a shield-shaped hole in the door that would once have housed a lock, a predecessor of the current arrangement.

Although the door has never been exposed to natural weathering, the lower ends of its boards show some signs of decay, especially on the north face, and part of the bottom inlaid ledge has been renewed (Fig. 28B). The lower hinge-strap and pintle have been replaced, doubtless as a consequence of being heavily rusted, and board 5 – the leading edge of the door – has suffered physical battering, consequent upon centuries of activity in the cellar. The decay may be attributed to prolonged low-level dampness in the building, potentially caused by flood-water emanating from the cloister garth. A

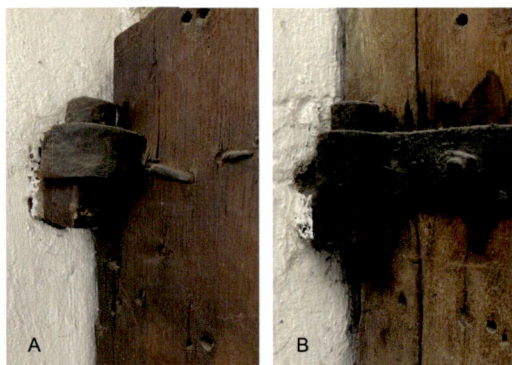

*Figure 41: Westminster Abbey, vestibule door. Hinge pintles: A, upper hinge; note also the clenched nails that attach the later hinge-strap on the north face. B, lower hinge (bolted), and replacement pintle. Author, © Dean and Chapter of Westminster*

huge volume of rainwater descends into the cloister from the surrounding roofs on all sides, and until Sir Christopher Wren installed an underground cistern and constructed an efficient drainage scheme at the beginning of the 18th century, flooding must have been a regular problem. The floor of the chapter house vestibule is at virtually the same level as the east cloister walk and any flooding in the latter will also have affected the former.[29]

Some timber repairs have been carried out in more recent times. The middle inlaid ledge (B) had a new end fitted where it is attached to board 5; this occurred at the time when there was a lock fitted in the large hole that is now covered by the iron place. Nearly one-third of the lowest ledge (A), where it is set into boards 1 and 2, has also been renewed, probably in the early 20th century, and slivers of modern timber have been inserted to repair damage to the abutting edges of boards 4 and 5 (Figs 25 and 27).

The date at which the hide coverings began to be removed cannot be established, but it was undoubtedly a gradual process. Over the course of centuries, the hides would have become tattered, and it is highly likely that the south side, at least, was progressively robbed by souvenir hunters and persons who found themselves in need of small pieces of hide for some purpose: *e.g.* to make a strap, to patch a damaged leather garment or shoe, or to make washers. Following the Dissolution of the Abbey in 1540, respect for its historic fabric and furnishings largely evaporated, and it was not until the reign of Queen Victoria that respect was regained.

Alongside both edges of the present upper hinge-strap are cut-marks on all five boards. They represent multiple slashes, hurriedly and carelessly made by a right-handed person, cutting away the remains of the hide (Figs 29B and 31). Traces of cut-marks are also visible at the bottom of the door, adjacent to the site where the 13th-century hinge was until Scott replaced it. More intriguing, is the scoring around the lower part of the C-scroll of the upper hinge (Figs 25, no. 12, and 39). Since the hinge was removed in the mid-13th century, the cut is unlikely to relate to stripping hide. The alternative explanation would be that it was a primary positioning mark to assist with fitting the hinge, before the hide was attached. This gains credence from the fact that initially the hinge was incorrectly fitted and had to be repositioned by 12 mm, as evidenced by the duplicate set of drilled fixing-holes for the C-scroll (Figs 38 and 39).

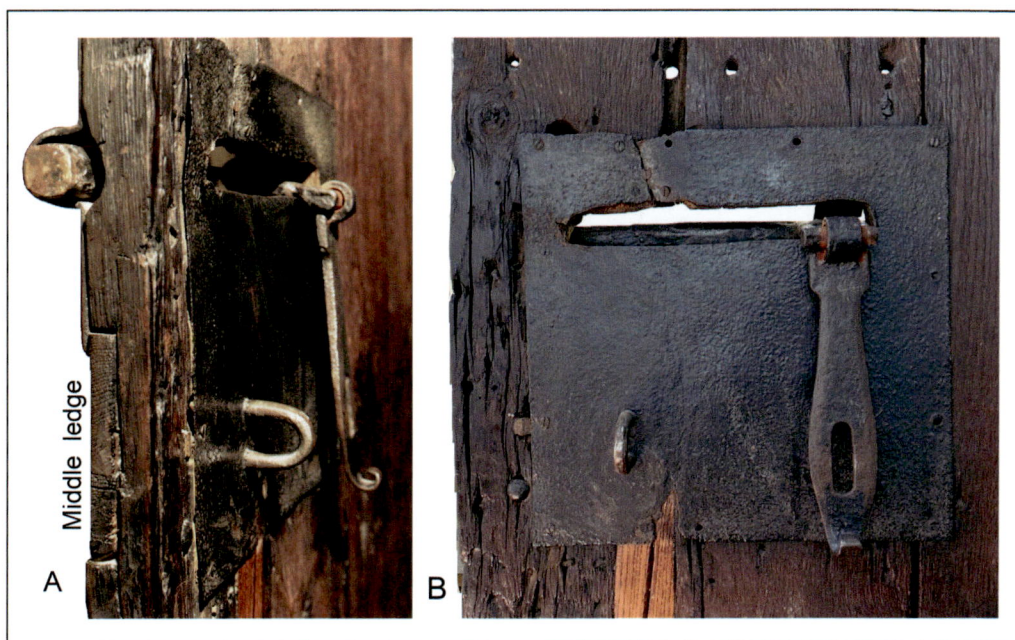

*Figure 42: Westminster Abbey, vestibule door. Detail of north face showing the partly medieval locking-bolt. The four small, drilled holes just above the plate relate to the lost primary locking device. Note also the housing for the middle ledge (B), exposed in the cut-back edge of the door. © Dean and Chapter of Westminster*

We have already noted that it was only the drilled holes for the C-scroll that were duplicated, not those associated with fixing the hinge-strap. This sheds light on an interesting aspect of the construction process confirming that the door was made on site, and not delivered as a finished object from a workshop at another location. The evidence reveals that the carpentry of the door was completed, followed by a trial hanging in its designated opening before either the hide or the metalwork was fitted. Eight holes were drilled for the C-scrolls, and these were bolted on to the door, but the split-curls were not nailed, and nor was the strap fixed. The door was then hung on its pintles to ascertain whether it fitted correctly. Evidently, it did not, and the upper hinge had to be repositioned; the scribed line under the scroll related to that correction. The temporary hinge fixings were then removed and the door taken back to the carpenters' workshop for the hides to be fitted and the ferramenta permanently attached.

Removal of the last remnants of hide from the door was a tidying-up operation, most likely undertaken in the mid-19th century, when the vestibule became the access route for visitors to the newly restored chapter house. Interestingly, the door was not subjected to mutilation by graffiti, in the way that parts of the interior of the Abbey,

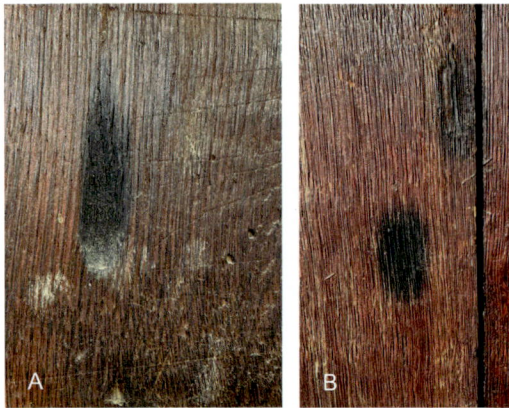

*Figure 43: Westminster Abbey, vestibule door. Candle-burns. A, tear-shaped apotropaic mark on the south face of board 5. The multiple knife slashes to the right relate to cutting off the hide. B, two burn-marks, probably accidental, on the north face of board 2. Author, © Dean and Chapter of Westminster*

and the Coronation Chair in particular, suffered. However, two black marks on the north face, and one on the south, of the door are candle-burns. The latter is tear-shaped in outline and displays the characteristics of an apotropaic mark, deliberately created inside the cellar. The other marks are amorphous and perhaps more likely to be the accidental consequence of a candle-stand in the vestibule being placed too close to the door (Fig. 43).[30]

In 2000–05, detailed archaeological recording was instigated in the Pyx Chamber and adjacent cellar, which included the vestibule door and its structural setting. A separate study of the entire monastic dormitory undercroft is being prepared for a future publication.

# 4

# A carpenter's study of the Westminster Abbey door

*Peter Massey and Paul Reed*[1]

The vestibule door is a rare example of mid-11th-century carpentry. By studying it in 2020, we established both how the timbers were produced and how the door was manufactured.[2] We have also determined the many types of tools used by the highly skilled carpenters who made it. The door has been identified as having originated from Edward the Confessor's abbey, and the felling of the timber has been dated by dendrochronology to between 1032 and 1064, making it the oldest scientifically dated door in Britain.[3]

A detailed measured and photographic survey carried out in 2005 included front and rear elevation drawings (Figs 22–25). The door has been reduced in both height and width and its original location is unknown. It was very likely made under the supervision of Teinfrith, Edward the Confessor's 'church-wright', at the time of building the Abbey.[4] The door was reused in its present position in the mid-13th century.

## Brief description of the door

The door consists of five vertical, parallel, oak boards varying in width from 22.5 cm to 39 cm. Its existing width is *c.* 1.27 m, and its height 1.98 m. There is an original central iron strap on the door, but the hinge-straps no longer survive, although the ghosted outline provided by their fixing-holes can be clearly seen (Fig. 39). The door is believed originally to have been *c.* 1.37 m wide by *c.* 2.75 m high, with a semicircular top (p. 59; Fig. 36B). The dendrochronology report by Miles and Bridge confirms that the timber was from oak trees grown locally near London.[5]

The boards are 40 mm thick with oak ledges housed, or let-in, so that they are flush with the surfaces of the door. There is a central recessed ledge on the front;

at the top of the reverse side of the door are the remains of the housing for a second ledge, and a third, almost complete, is close to the bottom on the same side (Figs 26–28). Most of the top ledge was cut away when the door was reduced in size to fit its present opening, leaving only part of the matrix (Fig. 29A). The flush ledges are slightly bowed along their length, and the surviving iron strap partly covers the central ledge (Figs 27 and 45).

The ledges are the main element that prevents the door from sagging. The closing edge is uneven where it has been reduced by *c.* 10 cm to fit the present opening. This was crudely done, most likely with an axe, and one end of the central ledge was exposed (Fig. 42A). There is clear evidence that the door was covered with hide on both sides. This is probably why the ledges were recessed, so that there was a smooth, flush finish to receive the covering before the strapwork was fixed.[6] The presence of the hide or leather covering on both faces suggests that it was a high-status door, and probably decorated. As further evidence that it was fully covered with leather, one can see knife score-marks on the boards on either side of the top hinge-strap, where the face leather has been removed, leaving a strip behind the strap (Figs 29B and 31).

## Introduction to Anglo-Saxon woodworking

Over more than half a century, archaeology has expanded our knowledge of Anglo-Saxon and Viking structures and the development of early woodworking and building technology, through new discoveries of complex Anglo-Saxon centres such as those at Cheddar (Som.), Yeavering (Northumb.), Cowdery's Down (Hants.), Bishopstone (E. Suss.) and Lyminge (Kent), as well as the Viking settlements at Dublin and York.[7] These sites provide good examples of advanced Anglo-Saxon and Viking building technology, although still not fully understood. However, they are also high-status royal or monastic centres, where there is evidence of highly skilled, sophisticated woodwork and setting out.

Archaeologists Damian Goodburn, an expert in Anglo-Saxon woodworking, and the late Richard Darrah both carried out experimental woodworking, replicating tool-marks on waterlogged timbers found on sites where Anglo-Saxon and Viking structures have been preserved.[8] Many timbers have been recovered from London riverfront sites and freeze-dried for display. Similarly, timbers are displayed at Jorvik Viking Centre in York and the National Museum of Ireland in Dublin. Stemming from Goodburn and Darrah's experimental woodworking, there are now reconstructed Anglo-Saxon buildings at Jarrow Hall (S. Tyneside), West Stow (Suff.), Charlton (Butser Farm, Hants.), the Weald and Downland Museum (W. Suss.), and the 'House of Wessex' at Long Wittenham (Oxon.).

These open-air museums are based on archaeological sites where ancient domestic buildings once stood, and their reconstructions used split and hewn timbers. They have earth-fast structural timber members for the walls and internal supports.[9] Some of the reconstructions followed the advice of Goodburn and Darrah, using three basic tools: a broad axe, a narrow axe and an auger, the evidence for these being the tool-marks studied on waterlogged Anglo-Saxon and Viking timbers. Other known tools – broad and narrow adzes, chisels, planes and saws – were ignored.[10] This wider range of tools is likely to have been used on royal palaces and monastic sites where the king and the Church could afford to employ the most skilled craftsmen with more tools at their disposal. Simple domestic buildings did not warrant highly skilled input, as Goodburn has demonstrated. The builders of earth-fast structures did not have knowledge of bracing and hence relied on fixing timbers in the ground to prevent a building from falling over.

A problem for carpenters making doors was how to prevent the boards from sagging. Green planks, whether split or sawn, would warp and shrink. Fixing elaborate ironwork to doors prevented the boards from warping but did not stop timbers from shrinking away from the straps, causing the door to sag through its own dead-weight. To overcome the problem, door-makers also used timber ledges to hold the planks tightly together and thereby prevent both sag and wind.[11] The vestibule door is an excellent example of this: the ledges were let into the face of the rebated boards, which were all fitted tightly together and had been seasoned to reduce subsequent shrinkage.

Another door of *c.* 1128 described by Geddes and Hewett, at Durham Cathedral, had square rebated planks like the vestibule door, but let-in tapered ledges. These were slid into dovetail-tapered recesses from the edge of the door and after two years, when the boards had shrunk, the ledges could be tapped further into the tapered recesses to tighten them up. Elmstead provides another example of this form of door construction (Figs 110–112). It was a clever design, except that the door could not be hung in a recess until the wedged ledges had been driven home, two years after it had been made, and the surplus remains of the wedge removed; only then could the door function on its hinges and close in the opening.[12] Other doors had different types of ledges, for example, the Hadstock door of *c.* 1075, where there are D-section ledges, fixed with nails and roves.[13] There, the planks were jointed with continuous, bevelled rebates. On some early doors, to prevent sagging, the jointing of the boards had counter-rebates, which must have been very time-consuming to make, using a chisel and plane (Figs 112, 118, 119 and 143).[14] However, the counter-rebate was effective in preventing sagging. It was not until *c.* 1200 that diagonal lattice-bracing was added to the backs of doors, making them more rigid.[15]

## How was the door made?

The boards of the door were converted from the log by sawing through-and-through. The orthodox view, held by Goodburn, claims that Anglo-Saxon woodworkers did not use saws as no timbers have been found to date that show saw-marks.[16] Goodburn also rightly states that all Anglo-Saxon woodworking was produced from

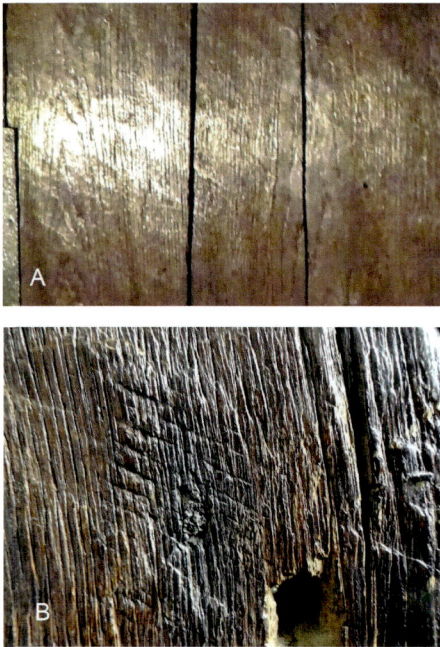

*Figure 44: Westminster Abbey, vestibule door. A, inner face. Diagonal marks left by the finishing tool, running across the boards, to remove saw-marks. B, outer face, showing saw-marks missed by the finishing tool. Paul Reed*

round logs squared up with broad axes, a process known as 'box-heart' conversion. Wet timbers studied by him did not have saw-marks, only axe-marks.[17] Other timbers studied by archaeologists were produced by splitting a log radially or tangentially and finishing it with a broad axe or adze. For the Anglo-Saxon woodworker it was necessary for the log to be straight-grained and knot-free to enable conversion, by splitting or hewing, to be carried out with ease.[18] However, the boards on the vestibule door have many knots, which would have been difficult to hew and near-impossible to split; hence it made sense to saw the logs to create planks. This could be an indication that knot-free trees for splitting were becoming scarce, necessitating the use of a saw.

Although Goodburn and others did not find saw-marks on the timbers that they studied, the evidence of the

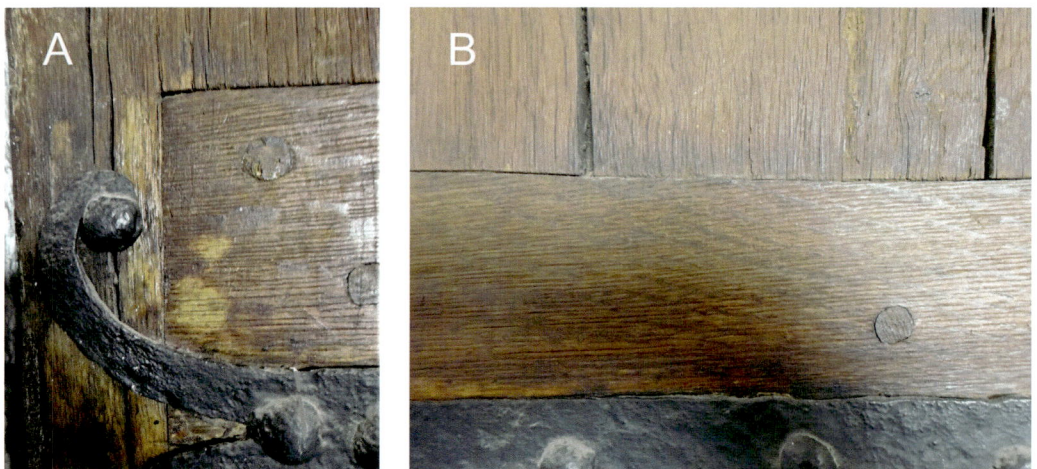

*Figure 45: Westminster Abbey, vestibule door. Outer face. Details showing how neatly the middle ledge (B) was let into the boards. Paul Reed*

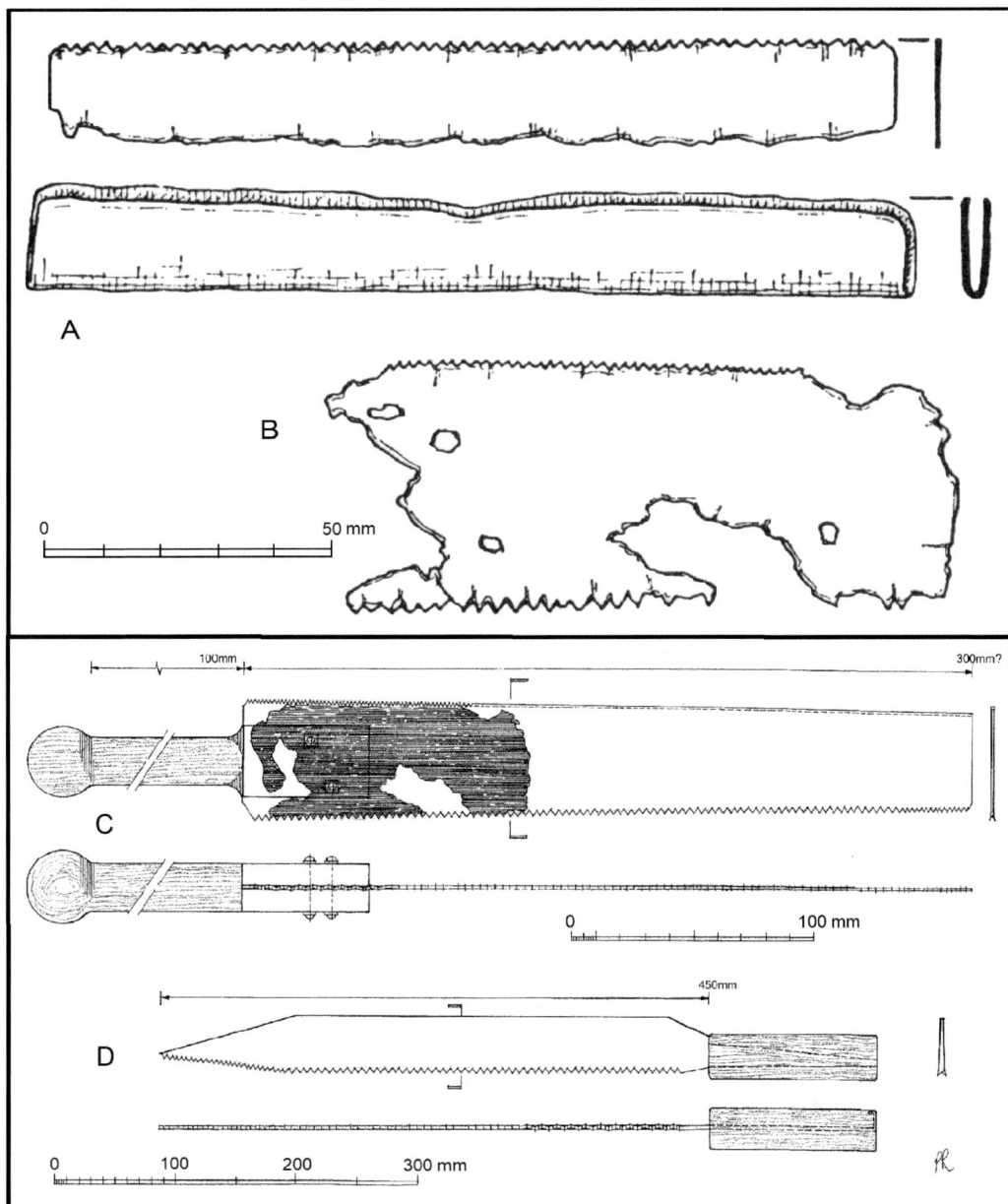

*Figure 46: Anglo-Saxon saw blades. A, Icklingham. B, Thetford. Reconstructions of saws. C, Thetford handsaw with double-edged blade. D, Viking handsaw from the Mästermyr hoard. A, B, After Wilson 1976, fig. 6.3; C, D, Paul Reed*

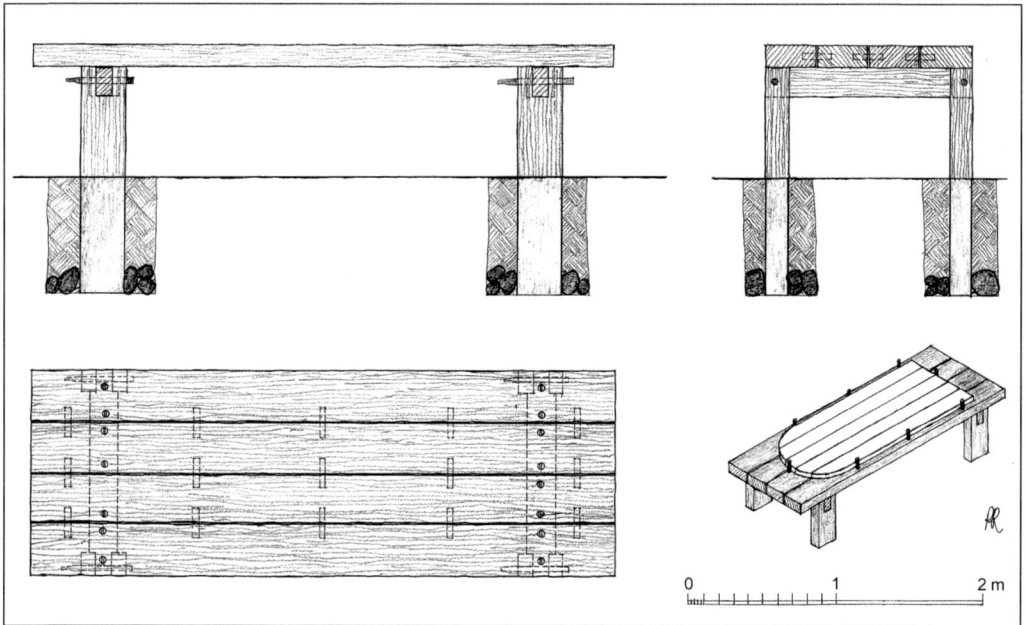

*Figure 47: Conjectural sketch of an Anglo-Saxon carpenter's ground-fast bench, with a door lying on it and wedged. Paul Reed*

vestibule door suggests that Anglo-Saxon and Viking carpenters did at times use saws. Bridge and Miles were the first to note saw-marks on the door (Fig. 44B),[19] and Hugh Harrison has observed saw-marks on the door at Staplehurst church (Kent), dated by Geddes to *c.* 1100.[20] This door was made approximately forty years after the vestibule door, and also has rebate-jointed boards, as well as D-section ledges like the Hadstock door.

Once the planks for the vestibule door had been sawn, they would have been stacked in the wood-yard on a level area, to be seasoned. Planks of this thickness would need to be 'in stick' for at least two years before being used. Anglo-Saxon woodworkers must have known about seasoning timber, as examination of the door shows that the ledges fit tightly in the boards with no shrinkage (Figs 28 and 45). However, later doors were still being made from green oak and were prone to shrinkage and sagging. Artefacts recovered through archaeology prove that Anglo-Saxon carpenters had knowledge of handsaws: *e.g.* the Thetford saw (Fig. 46). The evidence of the vestibule door, where the boards have clearly been sawn from the log, implies the use of a frame-saw for ripping along the grain.[21] This also provides proof that the Anglo-Saxon blacksmith could produce steel to make saw blades – over a metre long for a frame-saw – with teeth that could be 'set' to allow clearance of the blade, especially when cutting through a depth over 40 cm, as was the case with one of the boards in this door.

## Joining the boards together with rebates and dowels

After the planks had been seasoned, they would have needed straight, parallel edges, produced by flicking a chalk or a charcoal-dust base-line. Parallel lines could be achieved by using a lath or a gauge. The edges of the boards would be removed by sawing or by a broad axe, and truing-up using a plane. Straight edges were required to produce precise rebates like those found on the door. The boards would have been secured on a sturdy bench or trestles and were probably held in place with large removable pegs and wedges (Fig. 47). No Anglo-Saxon rebate-plane has been found to date but the evidence of the door implies that its maker used one. Only two Anglo-Saxon planes are known: one is made from bone and horn, and was recovered from a 6th-century grave at Sarre (Kent).[22] Bone planes have also been found in Friesland, Netherlands. The second plane is of wood and was found during excavations at Ebbsfleet (Kent) (Fig. 48A).[23] One of the authors (PM), recently made a replica of the Sarre plane (Figs 48B and 49), and we offer here a potential reconstruction of an Anglo-Saxon rebate-plane, modelled on the Ebbsfleet example (Fig. 50).

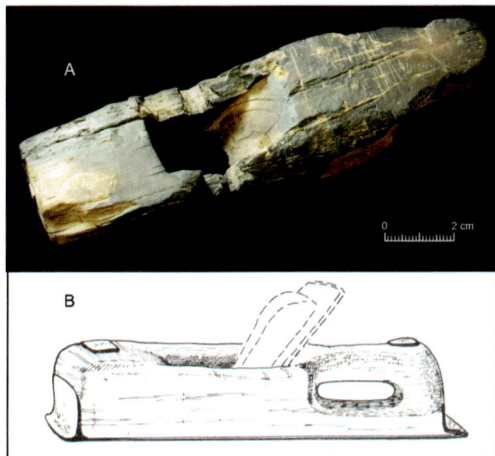

Figure 48: A, Anglo-Saxon plane from Ebbsfleet. B, small Anglo-Saxon plane found in a grave at Sarre. A, Robert J. Williams; B, after Wilson 1976, fig. 6.4

Figure 49: Replica of the Anglo-Saxon plane from Sarre, made by Peter Massey. Peter Massey

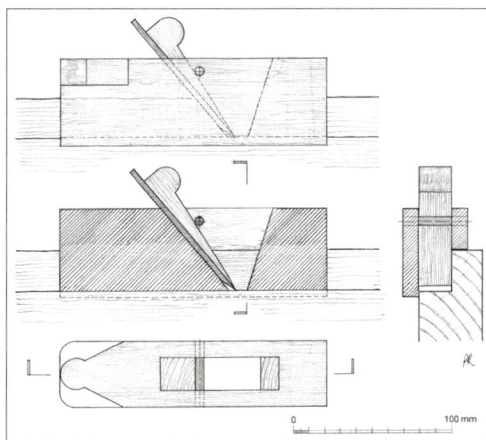

Figure 50: Conjectural reconstruction of an Anglo-Saxon rebate plane, modelled on the Ebbsfleet plane. Paul Reed

Figure 51: Hole made with a shell-auger to receive a locating peg in the rebated edge of the board. E. Massey

Figure 53: Shave made for the production of the replica door. Peter Massey

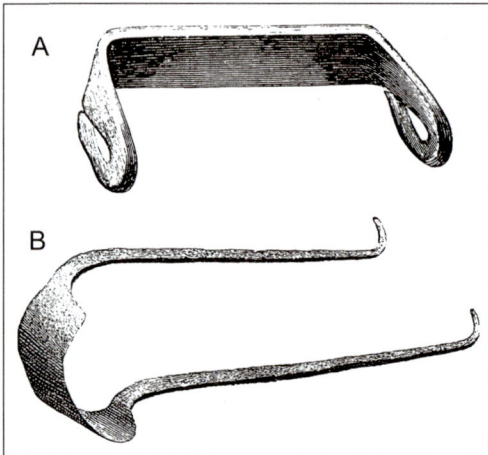

Figure 54: Viking split-socketed iron chisel from Skerne. Hull & East Riding Museum

Figure 52: Viking shaves found in Norwegian graves. Christensen 1982, fig. 18.2

In terms of the construction of the vestibule door, when the rebates were cut out on all the boards, holes were augered, four to a board (two on each edge), through the rebates to receive location pegs (Fig. 51).[24] These held the rebates together, and in line when the boards were assembled on the bench. With this done, the next step was to clean up both faces of the door by removing the saw-marks with a curved shave (Figs 52 and 53), or a plane with a curved blade. Such a blade would be similar in profile to the Viking chisel from Skerne (E. Yorks.) (Fig. 54).[25] Two shaves were found in the tool hoard excavated at Flixborough (N. Lincs.).[26] This process is visible where the distinctive, slightly concave tool-marks pass over adjacent boards, and is present on both sides of the vestibule door (Fig. 44A). These tool-marks have only survived through being protected for centuries by the hide covering.

### Letting in the ledges

The next stage would have been to let in the flush ledges. They are slightly bowed along their length, 14 cm wide at the ends, diminishing to 9 cm in the centre. The two surviving ledges are quarter-sawn or riven, and the medullary rays are present on the face. They were very precisely made from a flat board of well-seasoned oak, free from knots, 1.10 m long, 140 mm wide and 15 mm thick (Figs 26–28, 55 and 63). The ledges are about 15 cm shorter than the original width of the door. The long sides of the surviving central ledge are concave, creating dovetail ends that are square and neatly finished, possibly by a handsaw (Fig. 46). The centre ledge, being curved along its length, was probably set out with a straight-grained lath made into a bow, secured with a cord, and marked out along the edge of the lath with a sharp knife to produce a clear, curved line on the face of the ledge (Figs 55A and 61A). With the ledge timber securely fixed on edge, on the side of the bench or trestle, using the peg-holes in the ledge itself (Fig. 55B), the marked, curved score-lines were cut out, using a small hand side-axe, or a chisel and mallet, or a saw, to remove the waste (Figs 61B and 62A), and then finished with a draw-knife to leave a square edge (Figs 55B and 62B). The housed, or let-in, ledge on the bottom of the door has straight sides which could indicate that it has been replaced and is not original. This ledge also has a recent repair at one end (Fig. 28B).

The ledges would then have been laid on the boards and used as templates, their outlines scored with a knife, and the trenches to receive them made using a chisel (Fig. 55C). The fragment of the original matrix at the top of the door, exposed when it was shortened in the 13th century, provides clear evidence of marks made by a chisel with rounded corners and a sharp cutting blade, 40 mm wide (Fig. 58). Such chisels have been recovered from archaeological contexts. Hewett mentions that the housing was created 'with some kind of cutting gauge to cut the radii of the socket recess edges to the correct depth', assuming that the carpenter making the door had the room

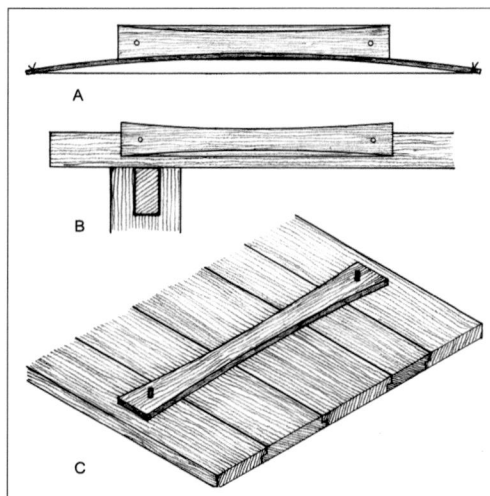

*Figure 55: Constructing and fitting a ledge. A, setting out the curvature of the long sides. B, cutting out. C, marking around the ledge on the assembled door planks. Paul Reed*

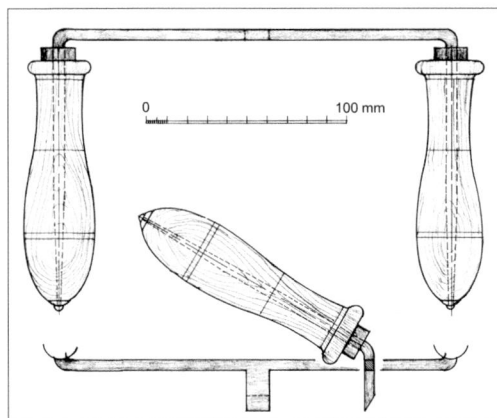

Figure 56: Conjectural reconstruction of a router, modelled on the Viking grooving-iron from the Mästermyr hoard. Paul Reed

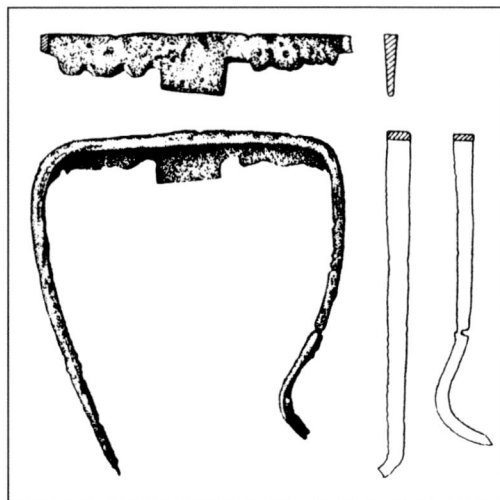

Figure 57: Grooving-iron, item no. 57, Mästermyr hoard. Nordiska Museet, Stockholm

and provision to make a large-radius cutting tool.[27] Hewett had obviously not seen the chisel-marks at the top of the door.

The recessed edges of the housing were formed using a chisel and a router (Figs 56 and 64). There is no evidence for an Anglo-Saxon router, but the Vikings used a grooving/moulding iron, similar in shape to a draw-knife, to make the base of the housing flat and consistent in depth (Fig. 57).[28] This can be seen at the top of the door where the ledge has been lost, leaving the matrix exposed (Figs 58 and 64B). The carpenter fitted the ledges into their matrices very precisely, and even today there is no gap between the ledge and the boards of the door (Figs 28 and 45). Once the central ledge was let in, the pegs were inserted to secure it to the boards.

The door was then turned over on the bench and the two remaining ledges fixed as described above. We could find no end-wedging to the pegs securing the ledges, as initially shown in Hewett's drawing.[29] In the case of ledge (A), all the pegs securing it pass right through the door; in ledge (B) just one peg is 'blind' and does not appear on the back face.[30] Since there was no fox-wedging, an adhesive may have been employed to ensure that the pegs did not come loose.[31] Glue was used to adhere leather to the faces of doors, and in the construction of shields.[32]

### The door-head radius

The most likely method of setting out the original head-radius of the door involved two nails and a piece of lath or string. One nail was tacked into the centre of the door and the other attached to the lath or string, so that a radius could be scored into the door planks. The surplus timber could have been removed with a narrow-bladed handsaw similar to that found at Mästermyr, Sweden (Fig. 46D).[33] Alternatively, the waste timber could have been removed with a very sharp side-axe, or a chisel and a mallet. Finishing the radius to the door-head would have been completed with a draw-knife.

### Finishing the door

Finally, with the ledges fixed in place on both the front and rear sides, the door was covered with hide and the ironwork attached thereafter. There is clear evidence that the surface of the tooled finish of the boards was rubbed over, across the grain, with a piece of coarse sandstone acting as an abrasive, to remove any highpoints (the same treatment a carpenter would apply today to smooth the surface of the timber with glass-paper). Once the carpenter had made a neat finish on the door, all the timber components fitted precisely. Could his intention have been for the door to be decorated without the need to cover it with leather?

*Figure 58: Westminster Abbey, vestibule door. Inner face. Remains of the lower edge of the matrix for the upper ledge, showing chisel marks. Paul Reed*

The Saxo-Norman north door at Hadstock has continuously splayed rebated planks, is flush on one side, was covered with hide or leather, and would have been painted red, like the vestibule door.[34] The door of *c.* 1100 at Staplehurst also has rebated boards, and the pattern of the medullary rays indicates that the boards were radially sawn.[35] This door, like the one at Hadstock, has D-ledges fixed with nails and roves. Once again, a plane and a rebate-plane must have been used. Another early door of *c.* 1140 at Kempley church (Glos.) has counter-rebated boards suggesting that a plane, a rebate-plane and a chisel must all have been used. Hewett drew the door with parallel boards each with three slip or loose tenons, whereas the door actually has tapered boards with four slip tenons as identified by Miles (Fig. 119).[36] Also, Hadstock has three Saxo-Norman timber windows with stiles with tenons on both ends, pegged into a mortised, arched head and a sill member.[37] This is another example of Anglo-Saxon carpentry/joinery where a bench must have been used.

It is the belief of the authors, that the Anglo-Saxon carpenter not only had a full kit of tools but was capable of making high-quality buildings, doors, windows and furniture. These tools, except for two types of axes and an auger, are not mentioned in archaeological reports because there has hitherto been no recognized archaeological evidence that such tools were used.[38] However, the Abbey vestibule door provides proof that other tools were skilfully employed by Anglo-Saxon carpenters, and woodworking

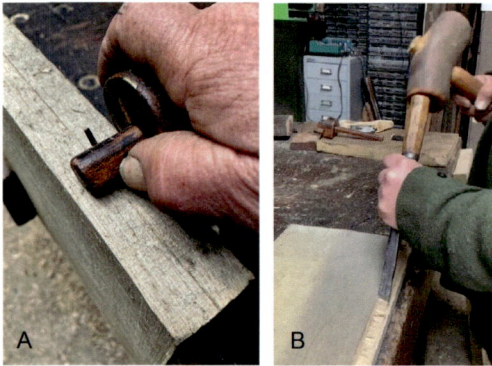

*Figure 59: A, setting out the rebates on the edges of the boards. B, cutting the rebates using a chisel. E. Massey*

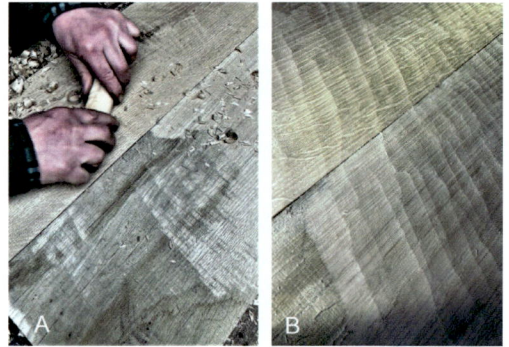

*Figure 60: Making the sample replica. A, shave being used to remove saw-marks on the faces of the boards. B, the finished result of shaving the sawn surface of the boards. E. Massey*

tools included axes, adzes, claw hammers, chisels,[39] augers, draw-knives and saws (Fig. 46).[40] Oslo Museum has Viking handsaws with 'set' teeth.[41] Hewett stated that St Cuthbert's coffin, made in 698, was constructed using a handsaw which left visible kerf-marks, and that possibly a chisel, router and rebate-plane were also employed.[42]

## Making a sample replica of the door

One of the authors (PM) made a sample replica of the vestibule door as a demonstration of how this was constructed, using a combination of modern and replica tools. A gauge was used to set out the rebate on the edges and the sides of the boards, taking the centre of the board as a measurement. The gauge is a simple tool that could have an adjustable fence, as here (Fig. 59A), or a fixed fence with a sharp nail as a marker. The gauge could also be used to measure the widths of the boards to ensure that they were parallel. When the rebates had been set out, the board was laid on the bench and held in place using pegs with wedges to prevent it from moving about while the rebates were being cut out with a chisel (Figs 47 and 59B). A rebate-plane would do the same job, and the authors are convinced that Anglo-Saxon plane-makers could have created a rebate-plane (Fig. 50). We assume that bench vices were not in use during this period.

The next stage was marking out the peg-holes on the boards to be joined and inserting the location pegs (Fig. 51). If set out accurately, the boards would fit perfectly, as seen on the original door. It was decided to cut the rebates first and then set out and sink the holes for the location pegs after the rebates were produced, as was done on the original door. However, it would be possible to sink the peg-holes first, before running out the rebates. Once the boards had been assembled and secured on the bench, they were cleaned up using a shave to remove the saw-marks on the surface (Fig. 60). For this task PM made a replica shave, adapting a modern one by re-forging the blade into a shape that would reproduce the historic toolmarks on the door (Fig. 53).

A suitable board was then selected for the ledge, free from knots, quartersawn and planed to the correct thickness throughout. The board was laid flat on the bench, and the concave radii set out (Fig. 55). This was a simple process using just a piece of string and a lath to make a bow. Drawing a large radius, as described by Hewett, was not necessary.[43] The ledge board was cut to length with square ends, using a handsaw. The bow was lined up on both ends of the board and the radius scored with a sharp knife (Fig. 61A). When both sides of the ledge had been radius-scored, the waste was removed using a hand side-axe (Fig. 61B) or a saw (Fig. 62A). The edges were finished with a draw-knife (Fig. 62B). Two holes were augered towards the ends of the ledge so that it could be securely held in place with temporary pegs while the boards of the door were scored with a sharp knife around the edges of the ledge, marking the area where it was to be let in. This had to be done very accurately so that the ledge would fit tightly (Fig. 63A). On the original door, one can see where the carpenter overscored with his knife.

The ledge was then detached from the door and the outline of the housing carefully cut, following the scored line, and the remaining waste removed

Figure 61: Making a ledge. A, setting out the radius with a bow and scoring with a knife. B, removing the waste with a side-axe. E. Massey

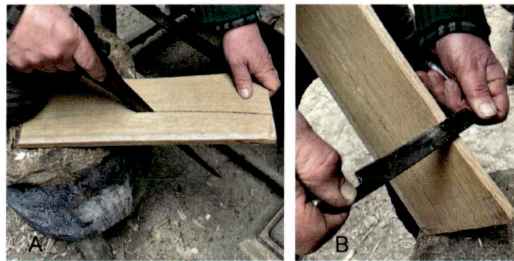

Figure 62: Cutting out a ledge. A, removing the waste with a saw. B, truing up the edges with a draw-knife. E. Massey

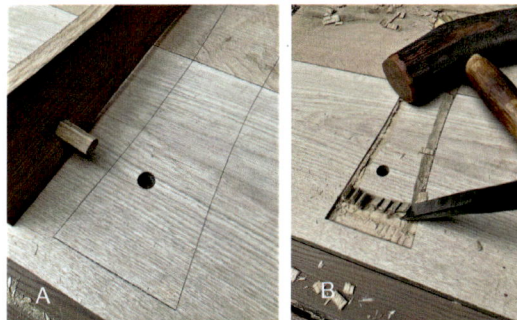

Figure 63: Creating the matrix for a ledge. A, the scored outline. B, cutting out the housing with a chisel. E. Massey

with the chisel (Fig. 63B). When all the waste had been cleared, a router was used to clean out the bottom of the housing to the required depth. PM used a granny-tooth router with a narrow blade (Fig. 64A). A sample of the ledge, a small piece of the

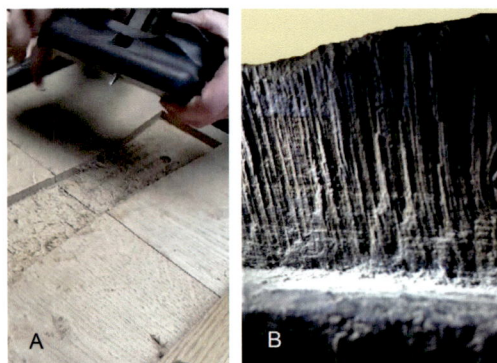

*Figure 64: Use of a router. A, routering out the housing for a ledge. B, horizontal marks left by a router in the partly surviving matrix for the upper ledge. A, E. Massey; B, Paul Reed*

*Figure 65: Completed replica sample of the vestibule door by Peter Massey. E. Massey*

same thickness, was used to check the depth of the housing. The marks of the router can be clearly seen on the original door (Fig. 64B). Finally, when letting in the ledge, it had to fit perfectly since it would have been impossible to remove for the purposes of adjustment without damaging the edges of the housing.

On the original door the central ledge was affixed to each of the outer boards (1 and 5) with two pegs and to each of the three interior boards with a single peg. The pegs are blind, meaning they do not go right through the board. The pegs remain tight in their holes, continuing to secure the ledge in its housing.

## Summary

The vestibule door is a singular example of Anglo-Saxon carpentry/joinery, revealing high-quality woodwork that is nearly 1,000 years old. The construction is uniquely different from other medieval doors found in the British Isles in that the ledges are flush on both sides. Our study suggests that its makers had access to a much wider range of tools than merely axes and augers. Since the Old English word *sage* meant 'saw', this alone suggests that Anglo-Saxon carpenters were familiar with saws.[44] The felling of the timber for the vestibule door has been dated by dendrochronology to c. 1032–64, making it the only surviving pre-Conquest door. It is clear that the vestibule door did not belong to the school of door-makers that used rove-fixed ledges, wedged ledges or counter-rebated planks. The only possible link could be the continuously rebated planks, but the method of fixing the ledges is different. The vestibule door is in a school of its own.

Anglo-Saxon timber buildings, doors, windows and furniture were built by carpenters having knowledge of setting out, using a line and plumb-bob, making mortice-and-tenon joints, lap-halving (possibly notched-lap/dovetail) joints, tongue-

and-groove joints and housing joints (as seen on St Cuthbert's coffin), and matched rebates using timber location pegs to hold timbers together (as found on the vestibule door). Carpenters were able to split or saw logs into planks and had the knowledge to season timber, square-up logs with hewing broad-axes, and make roof shingles.

The following tools were employed on the vestibule door: a plane (Figs 48 and 49), a rebate-plane (Fig. 50), a chisel and mallet, a router (Fig. 56), possibly a handsaw to trim the ends of the boards and the ledges (Fig. 46), a shave or plane to remove saw-marks from the face of the door (Figs 48 and 53), a draw-knife (Fig. 62B), an adze, a hand side-axe (Fig. 61B), probably a square for setting out boards and ledges, a flick-line, augers to make holes for pegs and clench-bolts, and finally the use of a bench (Fig. 47) or trestles in a workshop.[45]

For timber conversion, the woodman needed a felling axe, a lopping axe and a frame-saw that was used to saw the planks from the log, probably mounted on trestles, or over a saw-pit, or by adopting the see-saw method following the flick-line as a guide. The rake of the saw-marks on the door boards is about fifty-five degrees, as found on see-sawn timbers.[46] For lifting heavy logs or timbers, pulleys and ropes would have been required. Taken together this represents a substantial toolkit for a carpenter. We believe our study provides compelling evidence that Anglo-Saxon carpenters were highly skilled craftsmen, who were fully capable of precision setting-out and assembly, and making and adapting tools to do specific jobs, as demonstrated by PM in making the replica sample of the vestibule door (Fig. 65).

*Figure 66: Hadstock church. North porch, enclosing the Saxo-Norman doorway to the nave. Author*

imposts also with roll-mouldings and palmette-derived ('honeysuckle') ornament. The jambs of the opening are flanked by nook-shafts with cushion capitals that are also carved with palmette ornament (Fig. 68).[3] In this doorway hangs the notorious 'Dane-skin' door. It is not set into a conventional rebate, but closes flat against the slightly recessed internal face of the nave wall (Fig. 80).[4] This arrangement is found in other late Saxon churches. The arch and the door are complementary and there is no reason to doubt that they are other than contemporaneous. What is indeterminable is whether they have been bodily repositioned from a previous location on the north side of the church, possibly from the transept. The archaeology of the doorway cannot be properly studied because the wall is covered with medieval plaster.

## The legend of the 'Dane-skin'

St Botolph's Church is the most notorious of all English locations where the 'Dane-skin' legend has been perpetuated. Accounts of the doorway, as related by some antiquaries of the 18th and 19th centuries, have already been given, beginning with William Stukeley in 1724 (p. 13). However, the most florid of all descriptions of the Hadstock skin was published in 1847 by Richard Neville (later, Fourth Lord Braybrooke), a prominent English antiquary in the mid-19th century, and clearly a sceptic of the 'Dane-skin' story.[5] He resided at Audley End House (Ess.), and hence possessed good personal knowledge of the area that included Hadstock. Alluding to romantic antiquarianism, he wrote, somewhat amusingly:

> The marvellous-loving crowd will regard this sacred edifice [Hadstock Church] with peculiar veneration, as attached to the entrance [i.e. north door] was, for many centuries, an object qualified to gratify their taste, and really well worthy of comment. An outer covering, yellow and tough, handed down to a somewhat incredulous age, by forefathers much more confiding in such matters, as the actual skin of a Dane. Luckless individual, perhaps a stray one, captured and flayed alive by his mortal foes the Saxons, his hide being nailed to the church door, as a kind of warning scarecrow to those of his countrymen who might not wish to part with their upper garment.

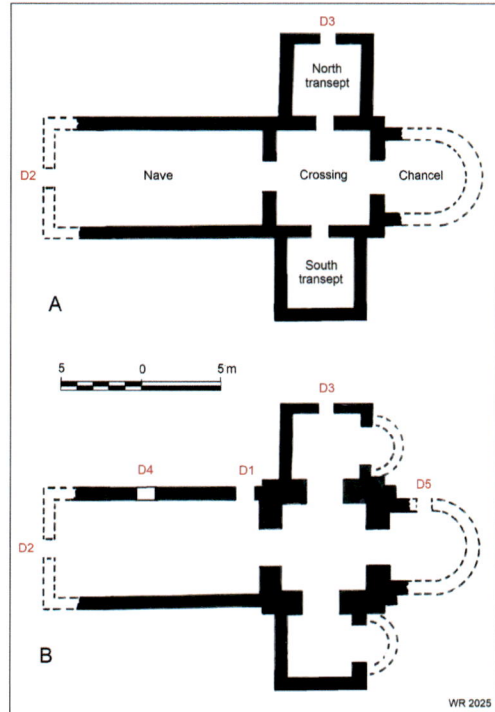

Figure 67: Hadstock church. Archaeological development plans of phases 1 and 2/3, showing the positions of the five known doors (D1-5). A, phase 1 (early 11th century); B, phase 3 (later 11th century), showing also the position of door 4 (phase 2 only). Author

Figure 68: Hadstock church, north doorway. West impost and nook-shaft capital with 'honeysuckle' ornament. Author

Of course, so pretty a tale finds ready credence, and not for worlds would we bring it into disrepute, merely insinuating that on the fragment of ancient portal, removed last year [1846] to make way for one at least weather tight, there is certainly *something* tawny in hue and coarse in substance; but whether the parentage so *flattering* to human vanity derives thence confirmation, we will not determine.

Part of the original wood-work now in my possession, bears on the surface nails and holes of a good size, shewing that care was taken that this poor persecuted piece of mortality should not after all give them the slip, and rejoin its kindred flesh. Through the kindness of the rector, Rev. C. Towneley, I am also provided with a piece of the hide of the robber, and can only further say, that this somewhat novel architectural decoration is uncommonly thick, and must have been thoroughly *tanned* both before and after its elevation.

A tradition of like character is affixed to the church of Copsford [*sic*], near Colchester, happily the only other instance of the recurrence of this strange ornament to the entrance of a place of worship – certainly but a dubious manner of illustrating the Christian precept, 'Forget and forgive'.

The name 'Dane' was deeply embedded in local folklore, as Neville noted: 'a little purple flower prevalent in the neighbourhood is called "Danes-blood"; and to the berries of the dwarf elder the villagers apply a similar appellation'. Dane was also a surname in northern Essex. At the same time as Neville was waxing lyrical, antiquarian scholars were beginning to take an interest in the possibility of confirming the identity of the

*Figure 69: Hadstock church, north door. Fragment of decorative iron scroll-work and a piece of hide. Courtesy of Saffron Walden Museum*

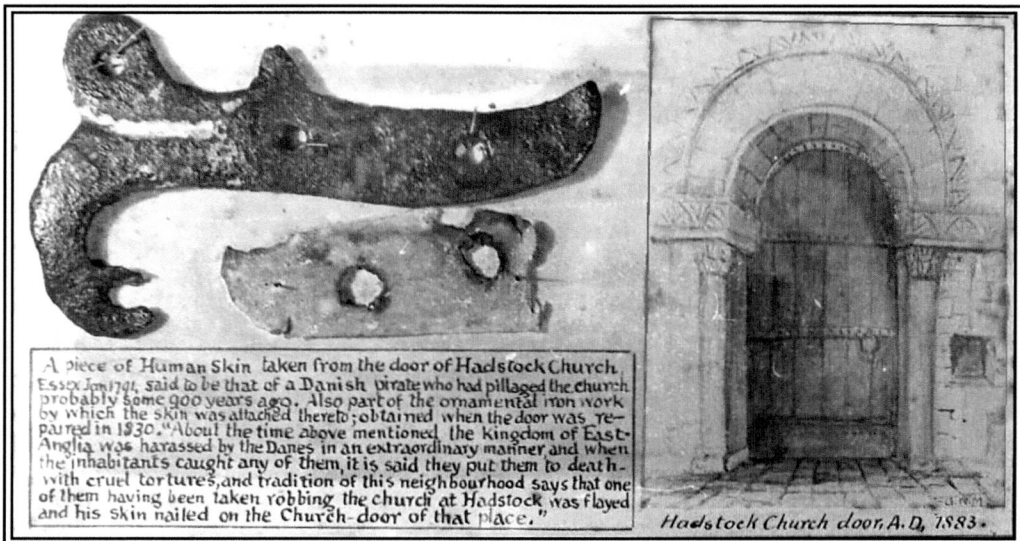

*Figure 70: Hadstock church, north door. Victorian museum display of fragments of iron and hide, together with a watercolour by G.N. Maynard, 1883. Courtesy of Jane Geddes*

skin by scientific examination. As we have seen, one of those was Albert Way (p. 28), who recorded that Sir Henry Englefield had 'laid before the Society of Antiquaries, in 1789, a plate of iron, taken by permission of the rector, from the door of Hadstock church, Essex, with a portion of skin, considered to be human, found under the iron'.[6] Those items do not appear to have survived.

Saffron Walden Museum possesses a piece of skin with one cleanly cut edge and two nail-holes ringed with rust staining; it is 7 cm long (Fig. 69). The outer face has evidently been much handled in the past and now presents a polished appearance, with no visible traces of paint. It is recorded in the museum register as having been 'taken from the door' in 1791. The museum also has a fragment of iron, being part of a small, barbed scroll with four nail-holes. A chisel-cut groove on the surface shows that the ironwork additionally bore incised decoration. The maximum dimensions of the fragment are 12 × 6.5 cm. The register records that the piece was 'obtained when the door was repaired in 1830'. The two items were accessioned in 1847, but the identity of the donor is uncertain.[7] A watercolour drawing of the door, made in 1883 by George Nathan Maynard, was displayed with the artefacts (Fig. 70).[8]

It is not difficult to appreciate why the posited Danish association is still so strongly championed locally. In the final stage of the Viking conquest of south-east England, a battle was fought in 1016 between the Saxon king Edmund Ironside and the army of the Danish king Cnut at *Assandun* (Ashdon/Hadstock in north-west Essex). Cnut was victorious, and the Anglo-Saxon royal succession came to an end.[9] He was also a Christian king and, as an act of thanksgiving for his victory over the English, Cnut erected a minster 'of stone and lime' at or near the site of the battle. The church was consecrated in 1020.[10]

Although the alleged Danish origin of the Hadstock hide has been strong for several centuries, an Essex antiquary, Miller Christy, challenged it robustly in 1925.[11] Curiously, no other writer had been moved to state the obvious:

> It is not easy to see how this fragment of skin can have been taken from a Dane, still less from a Danish king; for, in the battle fought at Hadstock [Ashdon], the Danes were victorious, and had no king killed, and there were no later fights with Danes on English soil.

Christy then offered his own wayward identification for the owner of the skin:

> In all the circumstances of the case, it is not altogether fantastic to suggest that the skin may be that of the traitor Eadric, retrieved from the City ditch or the Thames and nailed by Cnut's order on the door of his new *mynster*, erected on the very spot where the arch-traitor perpetrated his worst act of treachery, as a perpetual warning to all other traitors. Unfortunately, however, evidence is lacking.[12]

Ashdon and Hadstock are adjacent parishes, and the latter's early Romanesque cruciform church is customarily described by architectural historians as late Anglo-Saxon, or Saxo-Norman.[13] In addition, local historians in the 19th and 20th centuries have claimed the present building as Cnut's minster of 1020, a view that I initially supported. Later, discussions with colleagues persuaded me otherwise.[14] The distinctive architectural detail on the responds of the surviving crossing-arches, and the north doorway, are not paralleled by any known examples as early as 1020, but are more in line with work of *c.* 1070–80 (see further, p. 173).

## Antiquarian and modern study of the nave door (1)

Leaving aside macabre curiosity regarding the alleged nailing of human skins to church doors, serious antiquarian interest in both the stone portal and the door itself can be traced back to the mid-18th century. Several antiquaries living in Cambridge devoted great energy to visiting local churches, collecting information and writing notes. The earliest of which we have knowledge is the collection of albums compiled by William Cole (1714–82) that ran to over one hundred volumes.[15] Volume 35 of his 'Extraneous Parochial Antiquities', dated 1746, contains a bold, ink bird's-eye view of Hadstock church from the north-east, the earliest known illustration of the building (Fig. 71).[16] Cole was presumably the author of the drawing, which is particularly interesting for the fact that it shows the medieval chancel, before it was demolished in 1792. The two windows in the north wall were probably of the 14th century, but between them was a round-headed doorway of earlier date; it had chunky imposts and a door with two strap-hinges. This must surely have been 11th century.

Squeezed into a space beneath the drawing is the following secondary note:

> May 1. 1775. Mr. Essex left with me at Milton near Cambridge a very neat Draft of this North Door, with a MS Dissertation on Gothic Architecture, which as it is meant to be published I shall omit describing.

The reference is to the architect and builder James Essex (1722–84) who was also a Cambridge resident. Essex amassed a considerable collection of notes and drawings on antiquarian subjects, and visited Hadstock church sometime before May 1775. Regrettably, Essex's 'neat draft' of the north door has not survived amongst his or Cole's manuscripts.

Essex bequeathed his architectural collection to Thomas Kerrich (1748–1828), who also amassed voluminous notes and sketches. Their combined collections were substantial and Kerrich bequeathed them to the British Museum.[17] Both collectors acquired material from other parties, much of which is unattributed and undated: consequently, it is difficult to discover the authorship of many items.

Kerrich's assessment of the church reveals his archaeological approach to analyzing the building: 'The Church of Hadstock is a very ancient building but, having undergone many alterations since it was first built, it is difficult to determine what was its original form, there being at present but a small part of it now standing ...'.[18] He wrote this after the medieval chancel had been demolished in 1792. Kerrich's collection includes a pencil sketch of the north doorway at Hadstock and the artist initialled and dated it in very small characters, which appear to read: 'JAC Del. 1809'[19] (Fig. 72). A drawing of the south transept window is similarly signed.[20] These have variously been attributed to James Essex and Joseph Clarke, FSA (1802–95), a local gentleman-antiquary at Saffron Walden.[21] The initials are not those of either person, who did not have middle names.

Clarke produced drawings of excavations for Neville in the 1840s and 1850s, and his name occurs frequently in Essex publications. His association with Hadstock is confirmed by a drawing of the north doorway and notes that he made in 1850, initialled 'JᴴC' (Fig. 73).[22] These are the most fulsome notes written by any of the early antiquaries on the church; Clarke donated them to the Society of Antiquaries of London.[23] Following a brief architectural description of the building, he states:

*Figure 71: Hadstock church, from the north, 1746, showing the lost door 5 in the chancel. British Library, Add. 5836, fol. 17*

*Figure 72: Hadstock church, north doorway. Drawing in the Kerrich collection, dated 1809(?). Identity of the artist uncertain. British Library, Add. 6744, fol. 3*

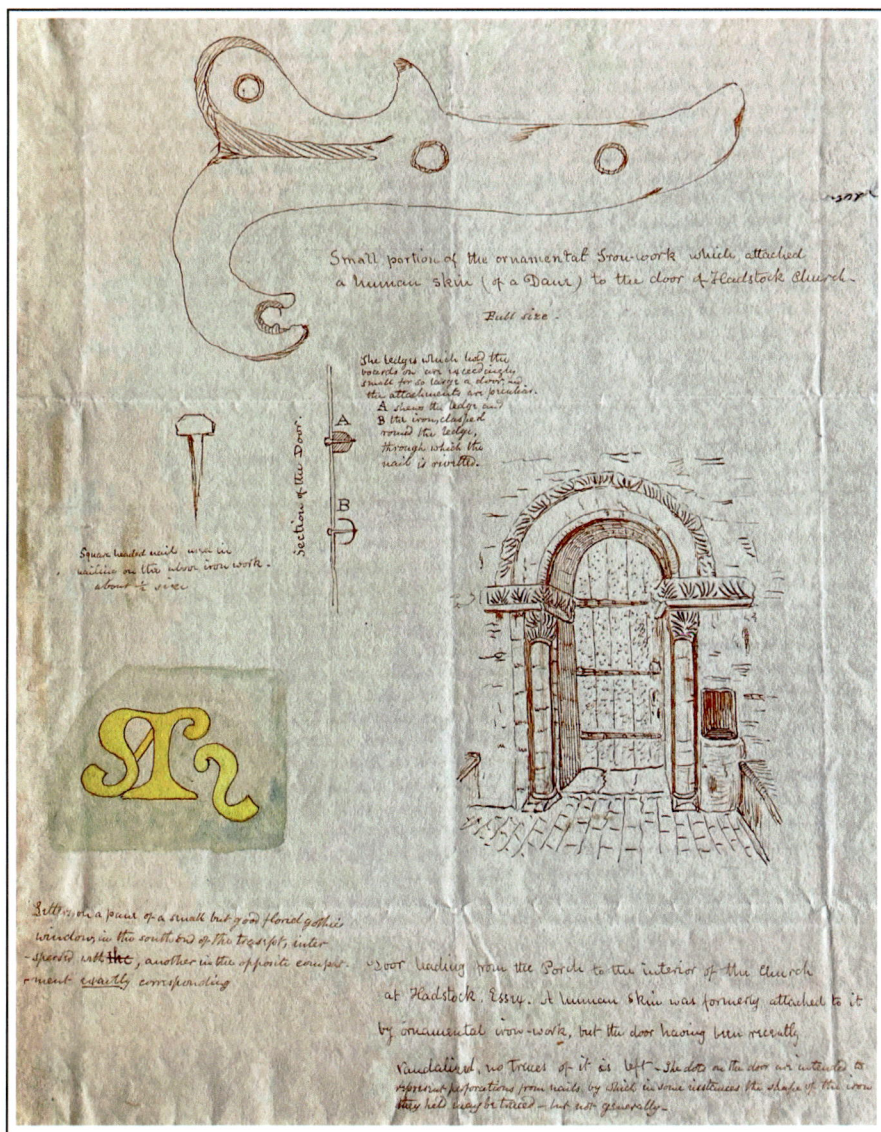

*Figure 73: Hadstock church. Notes (indistinct) and sketches by Joseph Clarke, 1850. Author, courtesy of Society of Antiquaries of London*

But the most popularly interesting part of the church and the talk of the whole country for ages is the door, and its skin covered with rude monumental ironwork. This door, is the door leading from the north porch into the church (see sketch) the speculation part of the question being whether the doorway with its piers and arch, is very early Norman or Saxon, if Saxon so also are the piers and one of the two arches of the transept, being evidently of the same period and workmanship [*i.e.* the 'honeysuckle' ornament]. The door itself is of oak and has lately been repaired, asphalted to darken its appearance, and entirely denuded of its remaining ornamental iron-work.

'The north door is much adorned with iron-work of an irregular form (probably regular when first attached) and underneath a sort of skin said to be that of a Danish king ... nailed on with many hundreds of (large square headed) nails, only small bits of it (which is extremely hard) remaining around the nails.' This was written in 1768.

There follows a citation from the Copford anecdote of 1690; and one from another quotation of 1815(?). Clarke continued:

... the curious, and I fear the antiquary is not altogether exempt, have been gradually pilfering pieces of the skin, and loosening the irons to get it from under them, until the Parish Clerk, an old man who has known it from his youth, tells me that but a very small portion had been left, and it had during all the years he had known it been gradually disappearing; and the door being out of repair, the whole was swept away (literally so, for the irons &c were swept away with the shavings and rubbish) leaving but the nail holes as a memento of the past, which is to be much regretted, the door being repaired as well upon the whole as you could expect a country village carpenter to do it, still in the neighbouring town of Saffron Walden there are several skillful workmen, quite alien to such a task, who would under the direction of an antiquary [have] kept up much of its original character.

Now I come [to] the interesting and controversial part of the question. Tradition from time immemorial has ascribed this skin to a Dane, a King, and probably for sacrilege, flayed alive and nailed to the church door, as a terrible example by the Saxons. Then it follows if the tradition is to be taken for anything, that the door must be Saxon, and if the more perishable door [is] of oak, there is nothing detrimental to the supposition, that the curious arch of the doorway, of more durable material stone, is Saxon also, and if these are Saxon then the piers with one arch of the transept are Saxon likewise.

But the only thing worth recording, appears to be that up to the early part of 1847 on the north door leading from the porch into the church of Hadstock in Essex were traces of a skin, conditionally said to be that of a Dane and there nailed under irregularly monumental iron work by the Saxons; at which time the door was repaired and all the then remaining traces of it obliterated.

The pen and ink illustrations comprise the following:

(i)   Full-size outline drawing of a portion of iron scrollwork taken from the church door. The maximum dimensions of the piece are 12 cm long by 6.5 cm wide (*cf.* Fig. 69).

(ii)  Example of a square-headed nail used in the construction of the door (approx. half-size).

(iii) Small sketch to illustrate how the ledges were fixed to the boards with nails and claw-like roves.

(iv)  Watercoloured drawing of a monogram in the stained glass of the south transept.

(v)   View of the north door and its setting, from within the porch.

The view of the door presents a conundrum: when was it drawn? The notes purport to date from 1850, and are certainly after 1846 because they detail the destruction that took place then. But this is not a view of the door in its 'restored' form, despite the legend stating:

> A human skin was formerly attached to it by ornamental iron-work, but the door having been recently vandalized, no trace of it is left. The dots on the door are intended to represent perforations from nails by which in some instances the shape of the iron they held may be traced – but not generally.

The door can hardly have been drawn by Clarke on the spot because the replacement iron hoop has been omitted and the three strap hinges are shown as markedly tapered along their length, whereas they are parallel-sided. This drawing is not an archaeological record,

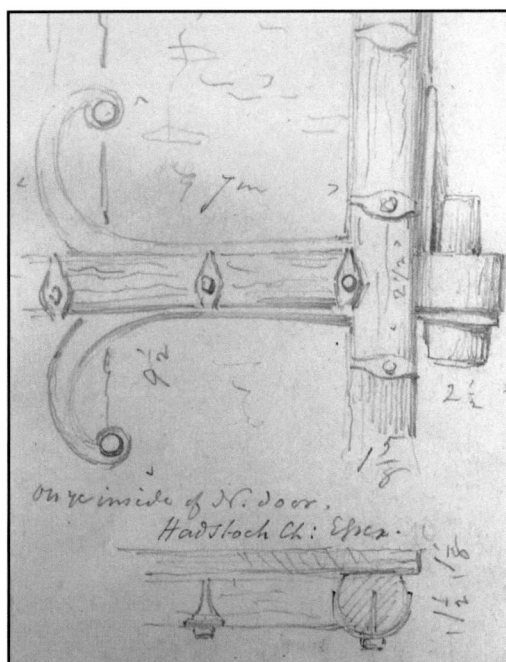

Figure 74: Hadstock church. Detail of the back-strap of a hinge on the north door. Drawing by J.C. Buckler. British Library, Add. 36433, fol. 601

but a concoction: the doorway has been closely copied (including errors) from the engraving of 1818 (Fig. 12), and the door itself was inserted from a faulty memory of its 'restored' appearance.

A passing reference in 1843 confirms the condition of the door immediately prior to its restoration:

> The skin of a Dane – some say a Danish king – (a portion of it is deposited in the Saffron Walden Museum) was till recently seen upon the entrance to Hadstock Church, covered with iron-work: the iron remains, but the skin has been taken away by degrees.[24]

Sometime before 1846, the architectural artist John Chessell Buckler (1770–1851) visited Hadstock and drew a detail of the back-strap of the middle door hinge (Fig. 74).[25]

The north portal is striking. In the first place, the remarkable palmette-derived decoration on the impost-blocks, arch moulding, nook-shaft capitals – as well as on what remains of the crossing arches inside the church – is without parallel in English Saxo-Norman architecture (Fig. 68). The door and its setting have been repeatedly examined by the present writer since 1973, including with Cecil Hewett in 1974 and with John Fletcher in 1975, when the latter made a record of its tree-ring sequence in

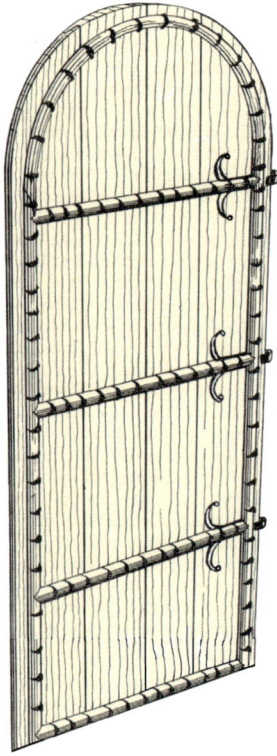

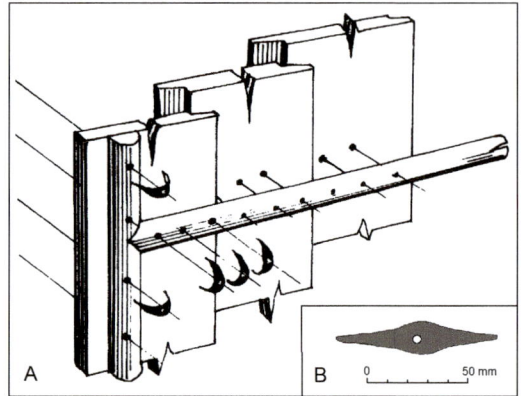

*Figure 76: Hadstock church, north door. A, exploded diagram illustrating its construction. B, sheet-metal cut-out for a clasping rove. A, Hewett 1980, fig. 20; B, author*

*Figure 75: Hadstock church. Perspective reconstruction of the inner face of the north door, showing its division into four panels coinciding with the positions of the hinges on the exterior. After Hewett 1980, fig. 19, amended*

the hope of dating it by dendrochronology (p. 123; Fig. 99). Hewett proclaimed that it must be pre-Norman: 'The earliest door that has survived in Essex is that rarity still hung in the north doorway of St Botolph's Church, at Hadstock. This is Anglo-Saxon and probably of the first half of the 11th century.'[26] He included sketches of the door and details of its construction in his first book on church carpentry. In 1978, he published a fuller account, accompanied by more refined drawings, including a perspective reconstruction and an exploded view of the complex jointing between the boards and the rounded ledges (Figs 75 and 76A).[27]

In the 1970s and 1980s Jane Geddes studied the door and published an account of its construction and ironwork.[28] It was not until 2004 that a fresh attempt was made to date the door by dendrochronology, taking micro-bores, which was successful (p. 125). At the same time, Adrian Gibson prepared the first detailed elevation drawings of both faces of the door, differentiating between surviving original carpentry and the heavy-handed Victorian restoration that replaced almost half of the boards (Figs 77 and 78). A seminar was held in the church on 8 July 2004, when all who had carried out research during the previous thirty years on the building – and the north doorway in particular – were invited to share their conclusions.[29] The present writer carried out a thorough re-examination in 2024–25.

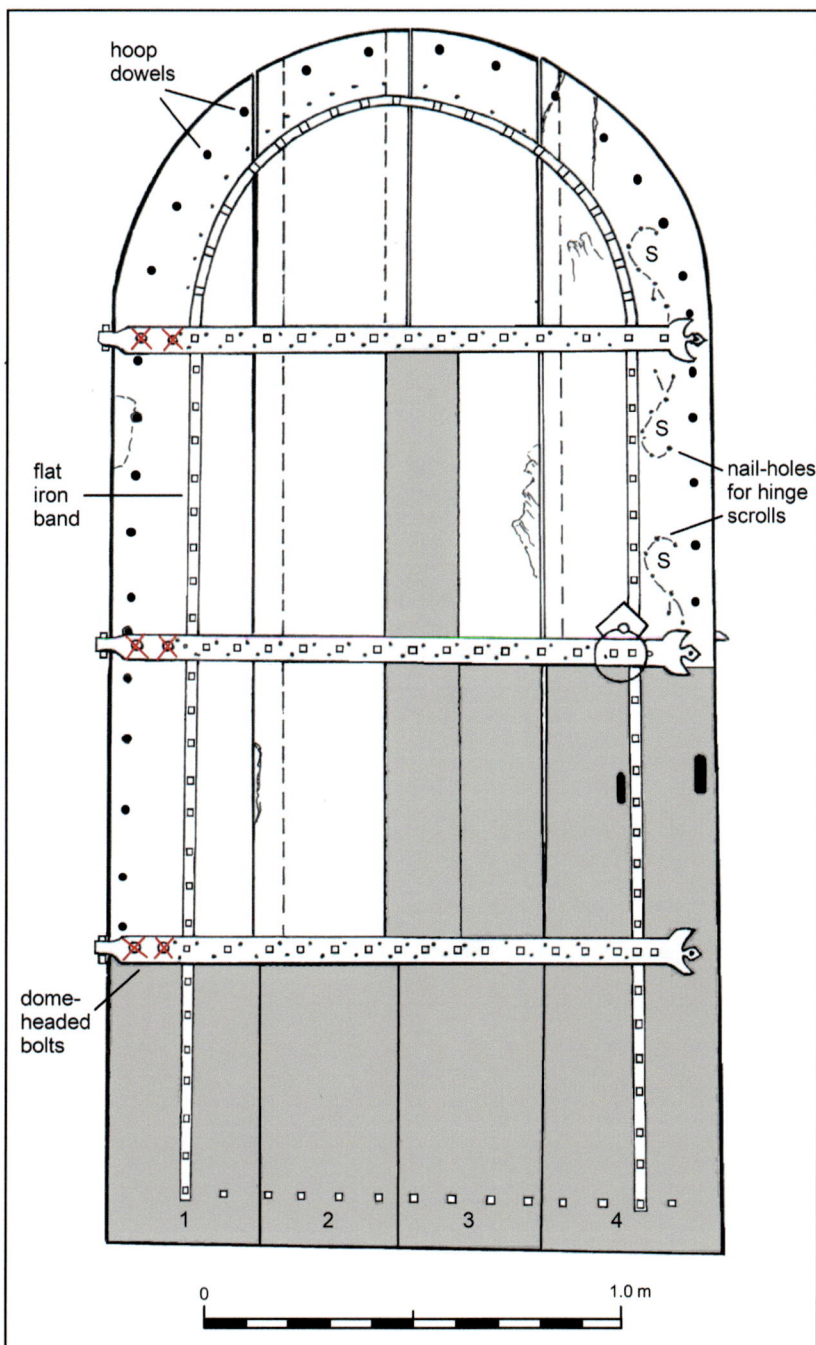

*Figure 77: Hadstock church, north door. Exterior elevation drawing. Grey shading indicates timbers renewed in 1846 and red crosses mark replacement bolts. Groups of nail-holes marked 'S' indicate where scrollwork decorated the ends of the hinges. Author, from a survey by Adrian Gibson, 2004*

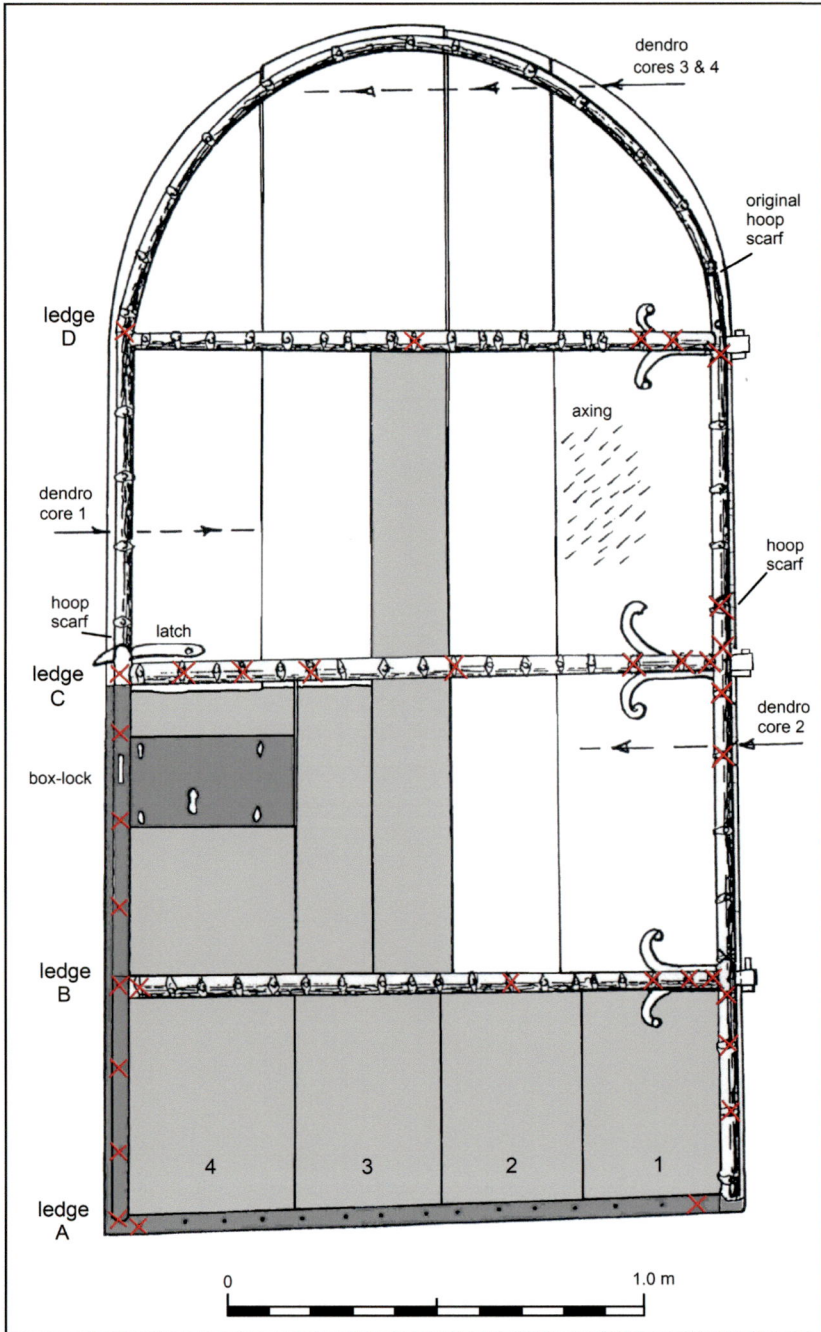

*Figure 78: Hadstock church, north door. Interior elevation drawing, showing the positions where cores 1–4 were taken for dendrochronological dating. Grey shading indicates timbers renewed in 1846 and red crosses mark replacement roves and bolts. Author, from a survey by Adrian Gibson, 2004*

There were two restorations of the door in the 19th century. One was a limited repair in 1830, and is recorded on a labelled piece of iron scrollwork in Saffron Walden Museum (Fig. 69), and the other was the thorough-going reconstruction in 1846, when much of the boarding was renewed (p. 101). The museum assembled a small display panel in the 1880s, which comprised the fragment of iron scrollwork, a piece of skin and a watercolour drawing of the restored doorway in 1883 (Fig. 70).[30] The display included a panel of text, which read:

> A piece of Human Skin taken from the door of Hadstock Church, Essex Jan. 1791, said to be that of a Danish pirate who had pillaged the church probably some 900 years ago. Also part of the ornamental iron work by which the skin was attached thereto; obtained when the door was repaired in 1830.
>
> 'About the time of the above mentioned the kingdom of East Anglia was harassed by the Danes in an extraordinary manner and when the inhabitants caught any of them, it is said they put them to death – with cruel tortures, and tradition of this neighbourhood says that one of them having been taken robbing the church at Hadstock was flayed and his skin nailed on the Church-door of that place.'[31]

## Form and construction of the north nave door

The door was constructed with four oak boards 30–32 mm thick and of varying widths, jointed with splayed rebates of 6 cm that are continuous for its full height.[32] It is 2.87 m high by 1.45 m wide, with a semicircular head (Fig. 79). Since the boards fit tightly together, with no significant gaps, the planks must have been seasoned before construction took place. The external face of the door is flat and was probably finished with a plane or shave, but any tool-marks have long since eroded away. Unlike the Westminster door, there are no drilled holes for fitting metalwork with clench-bolts. However, there may have been two clench-bolts in each hinge, where the front and back-straps grip board 1.[33]

The back of the door is framed with four ledges (A–D) and a bentwood hoop that borders the arched head, and extends down the sides for the full height (Fig. 80). The boards were not finished with a shave, even though they were visible from within the church. Areas of rough grain, knots and tool-marks are readily apparent, and on board 1 the scars left by diagonal axing are undisguised (Figs 82 and 83). No saw-marks have been observed.

The ledges and hoop are all formed of carefully rounded timbers, measuring on average 40–45 mm wide by 45 mm deep.[34] In cross-section, they are U-shaped, and bear prominent striations, revealing that a toothed blade was used to finish rounding the profile (Figs 81, 82 and 84). The timber for the hoop was skilfully bent, probably by steaming. It is attached to the boards by nails, driven in from the back of the door, but not emerging through the front. The length of the nails must be 60–65 mm (2½ ins). The ledges are nailed at intervals of up to 10 cm, and the hoop up to 15 cm. The arch of the hoop is made from a single piece of timber, scarf-jointed to the vertical

*Figure 79: Hadstock church, north door (open). Exterior, showing original hinges (restored 1846) and Victorian iron bands, dividing the door into four panels. Note the multitude of nail-holes in the boards where decorative ironwork was once fixed. Author*

Figure 80: Hadstock church, north door. Interior, showing rounded ledges and hoop-frame, dividing the door into four panels. The door is hung in a shallow recess in the wall-face. © Paul Ravenscroft

*Figure 81: Hadstock church, north door. Interior, showing detail of the back-strap and split-curl of the middle hinge, together with six replacement roves of 1846. Only the two roves on the far left are original. The short linear indentations on the boards mark the sites of clenched nails that secured decorative motifs on the exterior of the door. © Paul Ravenscroft*

*Figure 82: Hadstock church, north door. Detail of two undisturbed clasping roves on the hoop-frame and some on ledge C. The three marked with red dots are Victorian replacements (pipe-cleats). Note also the nail-holes in the edge of board 5. Author*

framing. More than half of the framing on the opening edge was renewed in 1846, along with ledge A.

Evidence preserved in the boards reveals an unexpected degree of sophistication involved in forming and attaching the hoop. First, a length of timber was cut and shaved to a U-shaped profile; then it was skilfully bent to the required radius. Next, pilot holes were drilled in the hoop at 15 cm intervals, for nailing; it was laid on the door and carefully positioned. Winged washers ('clasping roves') were slipped onto nails, before inserting them into the pilot holes. Each nail was lightly tapped with a hammer, so that its point dented the face of the door. The hoop was then lifted off and set aside, revealing the marked locations for the nails. A hole 5–6 mm in diameter was drilled through the door at each nail-point, and a flush dowel was driven into it. Finally, the hoop was repositioned on the door and its nails driven home, each penetrating the end-grain of its respective dowel. With the hoop secured, the wings of the 'rove' could be hammered down to clasp the timber.

Thirty of these plugged holes are detectable on the exterior of the door (Figs 77 and 85). The purpose of this sophistication is clear: the long, slender nails used to attach the hoop would not have responded well to the force required to drive them into the side-grain of the ledges and boards: they would be liable to bend. Driving into the end-grain of the plugs was much safer. We do not know whether the nails securing the four ledges were similarly driven into plugs: if they were, the evidence is obscured by the hinge-straps. Many roves were lost when the door was partly reconstructed in 1846, and their replacements were trapezoidal and less claw-like.[35]

Approximately 170 roves were required to fit the ledges and hoop; some are now Victorian replacements and many have lost part of their wings through rusting. Few complete original roves survive: they were c. 80 mm long and cut from thin iron plates (Fig. 76B). It has been supposed that the pointed wings of the roves clasped the ledges, to prevent the timber from splitting when the nail was hammered in, but that is untenable (p. 161; Figs 82 and 84).[36] The wings of each rove had to be hammered around the profile of the timber, and that could only be achieved after fitting. No evidence of splitting is apparent, confirming that pilot holes were drilled, or made with a gimlet.[37]

*Figure 83: Hadstock church, north door. Marks of diagonal axing on the back of board 1. Author*

It is unclear why so much of the original boarding was replaced in 1846.[38] While the boards below the bottom hinge were significantly decayed – as indicated in the 1809 sketch (Fig. 72) – the presence of the porch should have protected the timber at a higher level from the effects of weathering. Most curious is the renewal of a section of board 3 above the middle hinge-strap: nothing is shown on the 19th-century drawings – such as a viewing grille – to warrant this replacement.

*Figure 84: Hadstock church, north door. Detail of roves attaching ledge D to boards 3 and 4. The red dot indicates a replacement of 1846. Author*

*Figure 85: Hadstock church, north door. Pattern of seven nail-holes on the external face of board 4 revealing that the hinges terminated in pairs of S-scrolls (cf. Figs 91 and 139). The yellow rings mark the positions of two of the dowels to which the timber hoop on the inner face of the door was pinned. Author*

On the exterior, the surviving original boards display an enormous number of nail-holes, most of which fall within the confines of the iron hoop. While the majority were fixings for lost motifs, the replaced section of board 3 contains a smattering of nail-holes, all of which must relate to notices or decorations pinned to the door since 1846 (Fig. 86). Finally, there is an unexplained series of nail-holes around the outer edge of the door, running over the semicircular head and down both sides. The holes are spaced at 50–75 mm intervals and held substantial, square-shanked nails (Fig. 82). Board 1 has been split by them, just above the middle hinge. They evidently attached either an iron or thin timber band around the perimeter of the door, the most likely purpose of which was to secure the turned-over edge of the hide that covered the outer face.

### Hinges, roves and decorative ironwork

Antiquarian descriptions consistently state that the exterior of the door bore a heavy mantle of ironwork, which held the skin covering in place. Unfortunately, the two illustrations known to ante-date 1846, reveal little evidence of ironwork (1809 and 1818; Figs 12 and 72). The door is shown with three strap-hinges, and in the latter image all are depicted as having lost their outer ends.[39] In the earlier view, the hinge-straps extend across the full width of the door. However, this rudimentary sketch incorporates detail that is not otherwise graphically recorded. In the arch, above the uppermost hinge, an iron band in the form of a hoop with regular nailing is shown in the position where the present Victorian band has been affixed. Concentrically within that is a second, smaller hoop-band, lacking part of its right-hand side.

Close examination has revealed many of the nail-holes for both these hoops (Fig. 86). The upper part of the Victorian outer hoop was fixed *c.* 2 cm to the right of its predecessor, with the result that many of the primary nail-holes are visible.

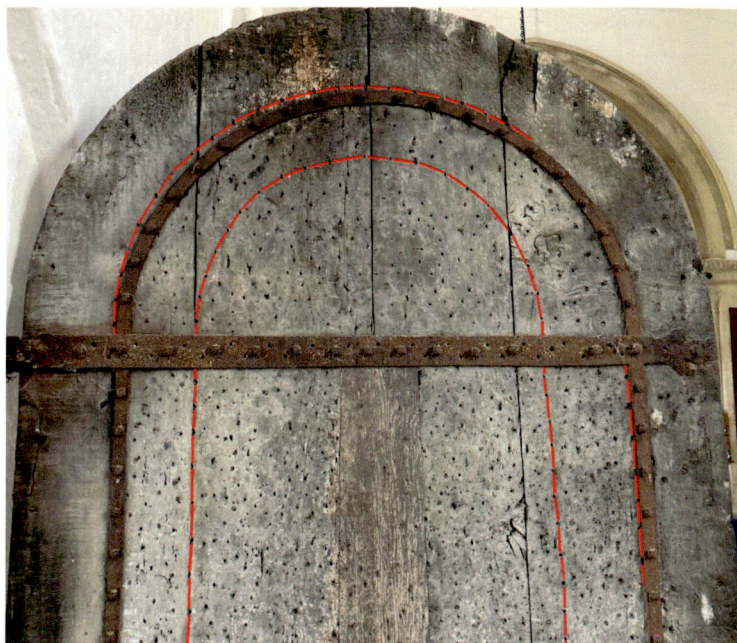

*Figure 86: Hadstock church, north door. Exterior, marked in red to show the centre-lines of the original outer and inner iron hoops, as revealed by the pattern of nail-holes in the boards. The Victorian outer hoop partly conceals the evidence for its medieval predecessor. Author*

*Figure 87: Hadstock church, north door. The three strap-hinges, as restored in 1846. A, top. B, middle. C, bottom. The arrows indicate the positions where new outer ends were welded onto the straps. Author*

Moreover, the downward continuation of the right side of the outer hoop can be traced across most of the face of the door between the upper and middle hinges, until the existing ironwork conceals the evidence. Nailing for the left side of the hoop is also concealed, but that for the iron bands for the inner hoop is fully visible within the arch and the panel between the upper and middle hinges. Below that, only the nail-holes for the left band of the hoop continue down to the lower hinge-strap, albeit interrupted by decay on the edge of board 2.

The hinges are not just fitted to the external face, but the straps wrap around the left-hand edge of the door, where they each incorporate a loop for hanging on a pintle set in the masonry of the jamb (Fig. 87). There is then a short return, or back-strap, on the inner face of the door, terminating in a split-curl (Figs 74 and 81).[40] Each hinge was made with two holes in the main strap and two more in corresponding positions in the back-strap, so that clench-bolts could pass through both plates and board 1, which was gripped between them. The straps are 60 mm wide by 6 mm thick and, in addition to the central row of fourteen or fifteen holes for nailing them to the boards, there are two flanking rows of smaller holes of a purely decorative nature. The rows are staggered, effectively forming a zig-zag design (Fig. 139A). These hinge-straps are unusual in being so densely peppered with small perforations.

All three hinges were removed from the door during the 1846 restoration, in order to heat them in a blacksmith's forge and make good the effects of decay. The straps were lengthened by skilfully welding on new strips of wrought iron, to which trifid terminals were added. The welds are barely perceptible, but the spacing of the rows of small holes is greater in the new work than in the original; the trifids have been cut out of a flat iron sheet (Fig. 77).[41] The hinge-straps most likely had splayed butts to which scrollwork was welded, similar to the surviving example on the west door (Fig. 91). Nail-holes in board 4 confirm the former presence of S-scrolls that were the only decorative ironwork on the front of the door, lying outside the confines of the larger hoop (Fig. 139A).[42]

When the door was made, the bentwood hoop on the rear was fitted after the hinges had been attached, but the ledges must have been nailed on first, since they initially provided the only means of holding the four boards together, unless temporary battens were affixed and later removed. However, the back-straps are oversailed by the right-hand ends of the ledges, the undersides of which had to be rebated by 6 mm, to accommodate the thickness of metal. The length of each rebate was 20 cm and it was cut before the ledge was nailed to the boards. The rebates were longer than was required, providing leeway for fitting and correctly aligning the hinges with the pintles that would already have been leaded into the masonry of the doorway.[43]

Next, came the attachment of the hinges, and that involved sliding the back-straps into the rebates under the ledges, with the split-curls springing out to either side (Fig. 81). The hinges were both bolted and nailed to the boards on the front of the door. Finally, the upright member of the hoop was fitted to the edge of board 1. Since this member had also to cross over the back-straps, housings 6 cm wide were cut into the framing timber at the three appropriate points.

Temporarily removing the hinges in 1846 was tricky, and the final stage of the construction process, just described, needed to be reversed: the framing timber on the edge of the door had to be prised up, but was not entirely removed, to allow the back-straps and their projecting split-curls to be drawn out of the rebates in the ledges. The nails securing the roves had to be extracted and, many of them, being firmly rusted into the oak, probably broke off, leaving their shanks embedded in board 1.

When repairs to the hinges were complete, they were refitted to the door and the hoop re-secured with a mixture of new nails and machine-made bolts. Each hinge was reattached to board 1 with two bolts with domed heads (Fig. 77).[44] Reinstating the straight section of the hoop involved replacing some roves. The bolts are not visible internally because they are covered by the timber hoop. A possible explanation is that the hoop was initially attached to the boards by pegs, and that the nails and roves were secondary. Some doors had their ledges attached by pegs alone (*e.g.* Rochester and Copford), others by nails and roves, or occasionally by a mixture of both (*e.g.* Stillingfleet). At Runhall (Norf.), Geddes noted: 'originally the door was held by three half-round ledges fixed by clasping roves. Each nail is hammered through a dowel peg in the ledge'.[45]

On the exterior of the door, the parallel band between the two iron hoops was filled with decorative motifs, as was the area within the inner hoop.[46] Curving rows of nail-holes indicate that scrollwork was included, and short runs of aligned nails possibly point to geometrical forms as well, but I have not attempted a speculative reconstruction of the patterns represented. A piece of iron taken from the door in the 19th century is evidently the terminal of a scroll, smaller in scale than those attached to the hinges (Figs 69 and 139C).

In the 1818 drawing, only the remains of the semicircular, outer iron band are dimly visible in the arch, and several areas of vertical shading represent the remains of the former skin covering the middle and upper reaches of the door (Fig. 88). A plain closing-ring is shown on the right-hand edge, along with several keyholes.[47] There is no sign of any surviving scrollwork, although some was most likely present at least until the repairs in 1830 (Fig. 85). Decorative motifs were nailed to the door, rather than clench-bolted, and the points of many clenched nails appeared on the interior, but few now survive. Others are represented by characteristic shallow indentations in the boards.

During restoration, a semicircular flat iron band, 28 mm wide, was fitted, mimicking the arch and continuing as a hoop down both edges of the door.[48] All the new ironwork was fixed with 18 mm square-headed and facetted nails (Fig. 86). As mentioned above, on board 4, immediately outside the iron hoop, three identical groups of seven nail-holes reveal the former presence of S-scrolls, being the terminals of two of the hinges (Figs 77 and 85). The third hinge was doubtless similarly embellished, but the evidence has been lost.

*Figure 88: Hadstock church. Sketch of the north door before restoration, showing the patchy survival of hide on the upper areas. The engraving has been tinted (buff) to highlight the scraps of hide; this depiction is probably more indicative than accurate in detail. Author, after Cromwell 1818*

### Fitting the hide

The door was covered with cow hide on the outer face alone, before any ironwork was fitted. A logistical issue arose, however, from the fact that the four ledges, and the semicircular element of the hoop, first needed to be attached to the boards in order to hold the components of the door together. Not being dowelled on their edges, the boards were completely loose. As noted above, three of the ledges are coincident with the positions of the hinges, and cut-outs had to be provided to accommodate the thickness of the back-straps. Also, one vertical element of the hoop had to be notched over the back-straps, and hence this could not be fitted until the hinges were in place.

The solution was to attach the four ledges, the bentwood hoop and its left-hand framing member, using nails, dowels and clasping roves (as described above), leaving the framing member off the hinge-edge. Since the ends of the three ledges that had to oversail the hinges were already rebated on their undersides there was provision for the back-straps to be slid into place, and secured, after the hide had been attached to the front of the door. Finally, the right-hand framing member could be notched over the back-straps and fixed with nails and clasping roves.

The hide was attached to the outer face, potentially using glue (as at Westminster) and the hinges positioned. They needed secure fixing to this large, heavy door: simply nailing them to the boards would have been inadequate, hence the provision of back-straps. Machine-made steel bolts were substituted for the original clench-bolts when the door was reassembled in 1846. Finally, the external iron banding and a multitude of decorative motifs were nailed on the hide-covered face (Fig. 86).

Only one piece of hide survives, which has been so intensively handled that no paint survives on its outer face (Fig. 69). However, traces of red colour are mentioned in at least one antiquarian account.

*Figure 89: Hadstock church, west doorway. Although contemporary with the 15th-century tower, the door is dressed with reused 11th-century ironwork. Author*

## Archaeology of the west tower door (2)

The doorway in the west wall of the 15th-century tower has a two-centred arch with hollow-chamfered orders, constructed mainly in clunch. In it hangs a second door of interest, somewhat crude in appearance, 2.42 m high by 1.38 m wide (Fig. 89).[49]

*Figure 90: Hadstock church, west door. Interior, showing cross-boarding and two of the hinge back-straps, both partly obscured by modern battens. Author*

*Figure 91: Hadstock church, west door. Remains of the scrolled terminal to the middle hinge; also an escutcheon plate with two keyholes and a slender closing-ring. Author*

The door comprises two layers of boarding. The outer consists of four vertical oak planks 20 mm thick, all cut from the same tree and exhibiting pronounced shakes. The boards are butt-jointed, without rebates or edge-dowelling. Internally, the door is backed with wide horizontal boards, original to its construction, but the narrow, chamfered battens covering the joints are relatively modern additions (Fig. 90). A 17th-century box-lock with scratch-moulded decoration has been fitted (inverted) to the back of the door.

Externally, two horizontal lines of nail-heads are present, together with a small closing-ring and an ancient iron plate with two keyholes (Fig. 91). The door carries additional ironwork that is clearly older than both it and the tower, but is closely related to the fittings on the 11th-century north door (1).[50] The ferramenta passed unstudied until the door was described by Geddes.[51] The three hinge-straps are 60 mm wide and *c.* 5 mm thick. At the outer end of the middle hinge are the remains of a double S-scroll with two barbs (Figs 91 and 139B). Each strap is pierced by three lines of small holes, the middle one of which houses the nails that fix the hinge to the door. These hinges are *en suite* with those on the north door, and the fragmentary

*Figure 92: Hadstock church, west door. Top hinge-strap and part of a narrow iron hoop reused from an earlier door with an arched head. Author*

S-scroll is a close match for the groups of nail-holes at the ends of the hinges on that door (Fig. 85). The scrolls are welded to the strap, and the small holes in the two outer rows are paired, not staggered like those on the north door.

The upper two hinges have back-straps that wrap around the edge of the door, and terminate in split-curls, exactly as on the north door. The back-straps are partly obscured by the semi-modern battens (Fig. 90). The lowest hinge has lost its back-strap. The top and bottom hinge-straps both now end in a splayed butt, the result of scrollwork having been broken off at the point of welding; doubtless, all three hinges were initially identical. They are too long to have been made for the present door, and must have been recycled here.

Nailed to the upper part of the door is an approximate semicircle of flat iron bar, with a diameter of *c*. 82 cm; the bar is 20 × 4 mm in section (Fig. 92). It is oddly asymmetrical since, on the left, the curve gives way to a straight section, indicating that this item was part of a hoop. The absence of nail-holes in the door confirms that there was no additional metalwork here, apart from a lock and a plain closing-ring (Fig. 91). The diameter of the hoop is too small to have been part of an edge-band of either the north or the putative west door. However, a clue to its use is provided by the sketch of 1809 that shows the remains of two concentric hoops in the arched head of the north door, and this fragment is commensurate with the small, inner one (Fig. 86).[52] This strap fragment cannot be derived from the north door, because it has almost certainly been attached to the west door since the 15th century.

The hinges and the hoop on the tower door are clearly reused, having been salvaged from an earlier door which was essentially similar to that in the north wall of the nave. The most likely location of the parent door was in the west end of the 11th-century nave. It is entirely logical that elements from the redundant door, displaced when the tower was erected, should have been used for the new west entrance, albeit crudely applied. The simple explanation is that the 11th-century north and west doors were of similar size and both fitted with identical ironwork: three perforated hinges with scrolls attached to their butt-ends, and back-straps terminating in split-curls; also flat edge-bands and concentric hoops in the arched door-heads.

# 6

# Rochester Cathedral: north-east transept staircase door

Samuel Pepys tells us in 1661 that the Norman west doors of the cathedral were covered with hide, but they were unfortunately destroyed long ago, and no samples of the material are known to survive. However, the cathedral possesses another Norman door that was not hide-covered, but is adorned with elaborate ironwork, comprising three large St Andrew's crosses in circles. Although the decorated face of the door has always been accessible, from within the stair-turret, it had been overlooked by architectural historians until 1989, when Tim Tatton-Brown drew attention to its existence, and arranged for elevation drawings to be prepared.[1] Known as Gundulf's door, it is the third oldest, scientifically-dated church door in Britain, and dendrochronology placed the felling of the timber in the bracket 1075–1107 (p. 125).[2] It is additionally interesting because – like other doors in this study – it carries the remains of medieval polychromy, painted not on hide, but directly on the boards.

## Carpentry

At the entrance to the north-east stair-turret of the north-eastern transept is an oak door, heavily decorated with ironwork on what was originally its outer face, but is now the reverse side (Fig. 93). The original inner face of the door has later been clad with horizontal boarding, totally obscuring details of its construction. The turret is part of the east end of the church, built c. 1200–15, but the door is clearly older and has been adapted for reuse in this location. In its original form, the door was narrower, square-topped, and composed of four vertical boards, measuring overall 1.89 m by 80 cm, although it has lost some of its width. An additional board (no. 5) was fitted to the left-hand edge and a segmental piece (no. 6) added to the top so that the door could be hung in a semicircular arched opening (Fig. 94). Plain strap-hinges with splayed butts were clumsily fixed to the outer face, in place of those that were originally on the back of the door. These might be replacements, but they are more likely to be the originals hinges, reused. The hanging edge of the door was reversed,[3] and two large oak box-locks were fitted to the present inner face in the 18th century, obscuring areas of decorative ironwork, including parts of the upper and middle St Andrew's crosses. The history of reuse displayed by this door and that in the vestibule at Westminster is remarkably similar. Moreover, in each case, the primary location of the door is unknown.

The archaeology of Gundulf's door was examined by Tim Tatton-Brown and Jane Geddes in 1989 (Fig. 94). In 2002, the door was lifted off its hinges and taken outside the cathedral, where it could be better studied and photographed, and cores drilled by Daniel Miles for dating by dendrochronology. The two box-locks, each fitted with four bolts and square nuts, were temporarily removed to reveal the full extent of the decorative ironwork (Fig. 93).

The Westminster door was double-sided, designed for internal use, and covered with cow hide on both faces; the small proportions of Rochester's door, and the lack of decay in its lower region, indicate that it too has always been internal. No evidence of a hide covering has been observed, and the lack of discernible gaps between the ferramenta and the boards implies that there never was an intervening layer. The size of the door in its primary form, and its flat head, potentially indicate that it gave access to a confined space, such as a stair-turret, passage or perhaps a sacristy associated with one of the lesser chapels in the eastern arm of Bishop Gundulf's cathedral. Additionally, the fact that its decorative ironwork so conspicuously recalls St Andrew – the church's patron saint – might indicate that the door was situated in a context of liturgical importance; this would tend to favour a chapel or sacristy.

In part, the construction of the door is not dissimilar to that at Westminster: the edges of the four boards are continuously rebated and joined together with secret dowels. On the back of the door were five timber ledges that appear to have been surface-fixed, rather than part-trenched into the boards. Their fixing scars cannot be seen without removing the later cladding on what is now the external face. However, a narrow strip of the back face of board 1 is exposed along the leading edge, where the later cross-boarding does not extend to the outer limit of the old door, but forms a rebate for closing into the masonry arch. If the five ledges had been trenched into the boards, the ends of all of them would be visible in this rebated area. It is therefore safe to conclude that the ledges were surface-fixed.

In her account of the door, Geddes drew attention to several small, unexplained holes in the front that must be clues to former fixtures on the back of the door.[4] The holes, that evidently housed nails (for clench-bolts?), fall on five horizontal

*Figure 93: (opposite) Rochester Cathedral, north-east transept staircase door. Original external face, now inside the stair passage. Two large box-locks were temporarily removed to expose the concealed ironwork (cf. Fig. 94). © John Crook*

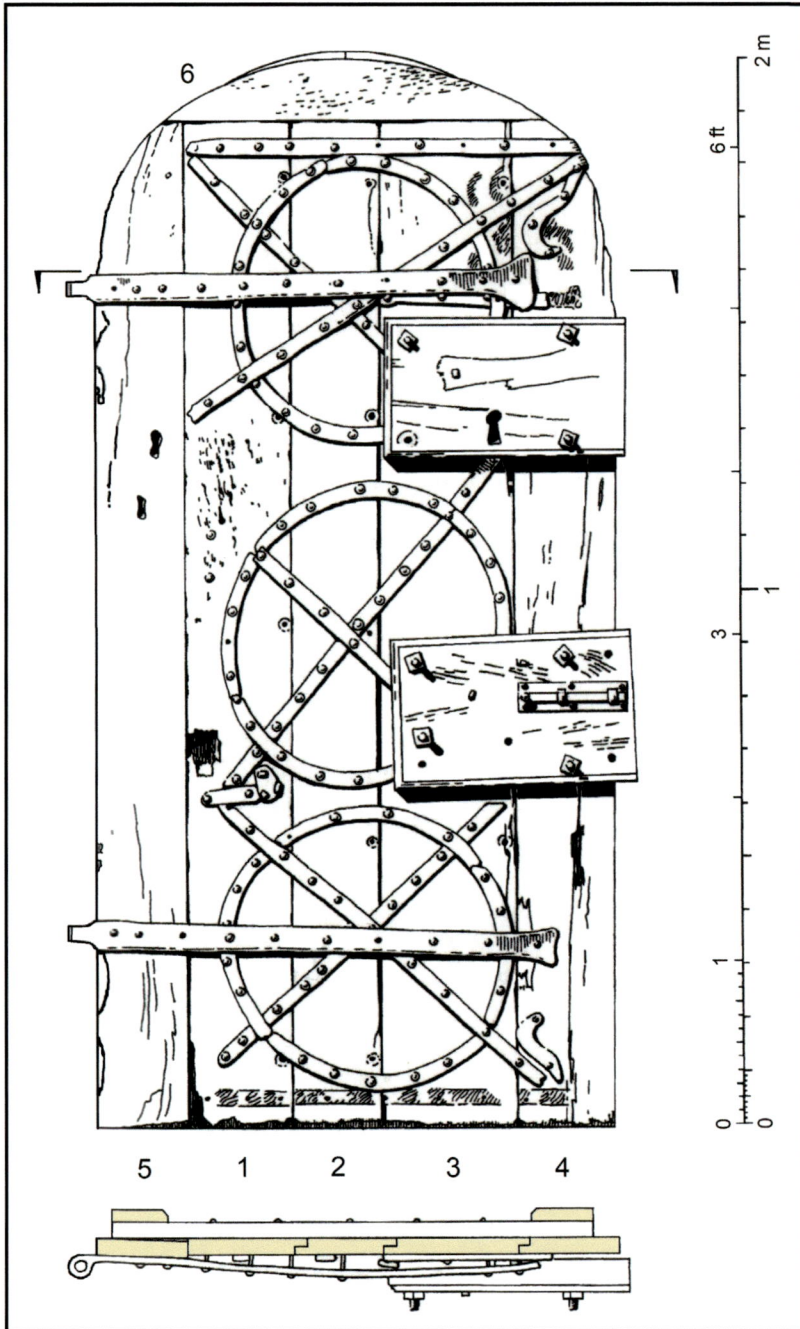

*Figure 94: Rochester Cathedral, staircase door. Original external elevation, with later box-locks and added boards (5 and 6); below, a cross-section at the level of the upper hinge, 1989. Drawn by Jill Atherton; courtesy of Tim Tatton-Brown*

lines, equally distanced (A–E). Five is the commonest number for ledges on medium-sized Norman doors. Not all the fixing-holes are readily visible, two having been obscured when the box-locks were fitted. However, at least three nail-holes have been recorded in each row. The spacing suggests that there may have been four or five fixings per ledge, but the outermost ones have been lost. It is a curious fact that all fifteen of the extant holes lie very close to, and on the left of, a joint between two boards. Consequently, the nails passed through the rebates, which is not a robust method of fixing the boards. Moreover, boards 1 and 4 appear to have no fixings near their respective outer edges, which would be essential to ensure that they were firmly anchored to the ledges.

Board 1 has very few surplus nail-holes, but there are two that might possibly have been fixings for ledges D and E. Even so, they represent smaller nails than the others used to secure the ledges.

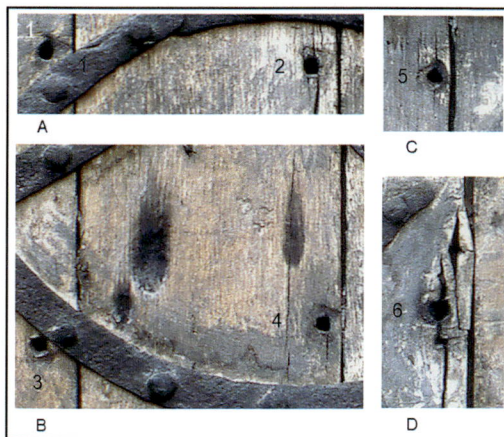

*Figure 95: Rochester Cathedral, staircase door. Details of indentations (nos 1–6) made by the nail-heads and roves that attached ledges to the boards. A, ledge E, boards 1 and 2. B, ledge D, boards 1 and 2. Note also three taper-burns. C, ledge B, board 2. D, ledge B, board 3; split caused by a nail. Hole nos 1, 2, 3 and 6 show impressions of square-headed nails or roves; no. 4 is round with one flat facet; no. 5 is round and small. © John Crook*

Locating possible nail-holes for fixing the outer edge of board 4 is more difficult: there are numerous scars close to the edge of this board, throughout its full height, mostly relating to the loss of decorative ironwork. Although nail-holes are identifiable that align with all five rows, they are again of smaller size, and could be fortuitous.

Most of the fifteen nails confidently assignable to fixing the ledges to the boards exhibit two characteristics: first, the nails have been withdrawn, leaving well defined square holes in the boards; secondly, the nails were driven into the boards with some force, creating circular indentations that reflect the dimensions of their heads (Fig. 95).

## Ironwork

The decorative ironwork principally comprises three St Andrew's crosses set in circles, all made from thin pieces of flat iron bar, fixed to the boards with multiple nails. Each circle was formed by joining two semicircles together. Interestingly, one of the bars forming each cross was nailed to the door before the circle was attached, and the second bar was added afterwards, so that it overlay the circle.

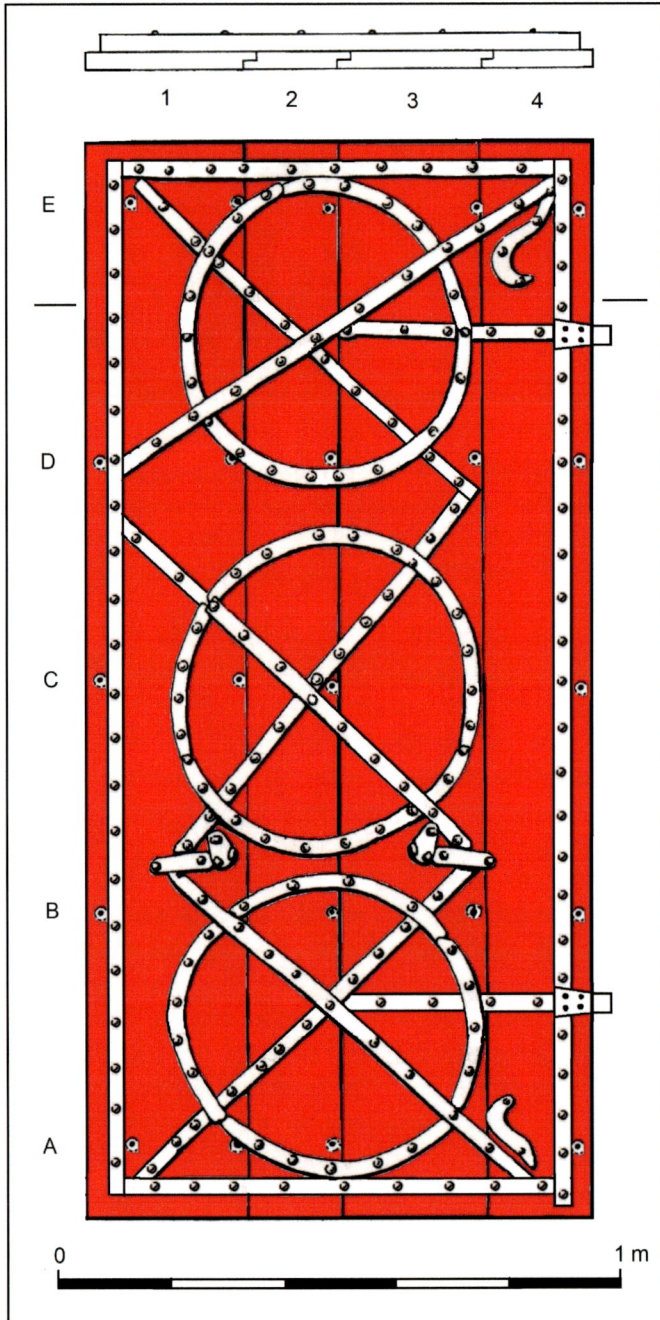

*Figure 96: Rochester Cathedral, staircase door. Reconstruction of the original appearance of the exterior. Five rows of holes where clench-bolts formerly attached ledges to the rear of the door are indicated by the letters A–E. Author*

It has been noted by Geddes and Miles that there were edge-bands along the top, bottom and right side of the door, but only the band at the top survives today.[5] The ghost of the bottom band is clearly preserved, together with the holes for the nails with which it was attached. The evidence for an edge-band on the right is less obvious, being confused by the multiplicity of nail-holes extending along the full height of the board.[6] Nevertheless, sections of the edge-band are clearly ghosted. Geddes also noted that partly underlying the top hinge is a narrow, horizontal strip of wrought iron, ending with an ogival lobe; it appears to serve no purpose, and nail-holes alongside the bottom hinge indicate that a comparable strip has been lost from there too. In both cases, the square ends of these strips abutted the back-straps of the original hinges when they were mounted on the rear face of the door, as shown in the reconstruction (Fig. 96).

These anomalous strips were primary because both are overlain by the St Andrew's cross motifs, and scarring confirms that they also continued further towards the right-hand edge of the door,

*Figure 97: Rochester Cathedral, staircase door. Detail of the top right-hand corner, showing the jointing and lapping of the decorative iron banding, white paint under the upper box-lock (temporarily removed) and patches of post-medieval grey paint. Key: A, primary white paint, base-coat. B, fragment of primary strap under later ironwork. C, strapwork of St Andrew's cross and circle (overlying B). D, iron billhook motif. E, top edge-band. F, repositioned hinge-strap, taken from the back of the door. G, grey paint. © John Crook*

stopping where they met the back-straps. An exactly comparable short length of narrow strap, terminating in an ogival lobe, is preserved on the south door at Pitsford church (Northants.).[7] It too lies alongside the lower (replaced) strap-hinge. These ogival-ended strips are too flimsy to be parts of hinges, but are decorative strapwork that abutted them.

The present hinge-straps are plain, and their butts are markedly splayed; they have been crudely placed over the upper and lower St Andrew's crosses and are fixed with multiple nails. Indications of the positions of the primary hinges are discernible; they were fixed to the back of the door, but there is no evidence for lines of clenched nails projecting through the front face, as might have been expected. However, this can be explained by the fact that when the hinges were relocated to the front face, their positions relative to the height of the door were unchanged; hence the primary nail scars remain hidden from view behind the repositioned straps. This implies that the reversed door was intended to hang on the same pintles as previously. The hinge positions do not synchronize with the ledges, but lie midway between A and B, and D and E, respectively (Fig. 96).

Moreover, it is noticeable that on board 4, where the present straps stop well short of the right-hand edge of the door, there is a confusion of nail-holes and areas of damage to the surface of the timber. This is consistent with the previous hinges, fitted on the back face, each having a short length of metal wrapping around and gripping the edge of the door, as a back-strap or plate.

Although overlooked by previous commentators, the asymmetry in the layout of the ironwork on the front of the door is striking, when the design so obviously calls for approximate symmetry. The three circles containing the St Andrew's crosses are not centred on the vertical axis, but confined to boards 1–3. They do not impinge upon board 4, the only surviving decoration on which is a small iron scroll – having the appearance of a billhook – at the upper and lower corners (Fig. 97). Nail-holes point to the loss of other small metal features here, but their design is irrecoverable. The evidence for an edge-band on board 4, just described, confirms that the original right-hand edge of the door is preserved.

There was no room for a matching band on the left edge, unless the door has been reduced slightly in width; instinctively, that might seem unlikely, because an extra board (no. 5) was added sometime, to widen the door. However, it is noticeable that the joint between boards 5 and 1 is perfect, with no sign of the casual damage that one would expect to see on the arrises of the opening edge of a door. Board 1 was almost certainly reduced in width by at least 5 cm, removing the scar evidence for an iron edge-band, and accentuating the asymmetrical layout of the decoration. Furthermore, if the hinges are primary, and recycled – as argued above – they are too long for a door that was only 80 cm in width, but would comfortably fit one of 85–90 cm. It is therefore maintained that when board 5 was added, the damaged primary edge of board 1 was trimmed and planed, to effect a tight joint.[8] The reconstruction offered here allows the edge-band to be completed, which is aesthetically essential for the design to make sense, but it does not introduce perfect symmetry to the placing of the St Andrew's crosses: to do so would require the hypothetical widening of board 1 to 35 cm (Fig. 96). That is certainly possible, but evidence is lacking.

## Painted decoration

Since this was an internal door, any hide covering should have survived well and, even if it was stripped when the door was widened and rehung, there should still have been pieces trapped under the ironwork. But none has been reported in the past, and a fresh examination failed to reveal any evidence for a covering. Moreover, residual traces of two phases of painting – directly on the timber – are readily apparent.

This important aspect of the Rochester door merits discussion. Geddes and others have noted the remains of dark blue-grey paint at the top right corner (Fig. 97), but there is also an extensive area within and adjacent to the lowest St Andrew's cross, and numerous small patches of the same colour elsewhere. Since this is not present under either of the box-locks, it is evident that the whole door and its ironwork were given a thin coat of grey paint in the 18th or early 19th century. Surface markings are compatible with the majority of this overpainting having been casually scraped off at a later date. The ironwork has rusted to such an extent that nearly all traces of paint have been shed.

It is the presence of red paint in the middle of the door that deserves closer examination, since it is undoubtedly primary. When the two box-locks were temporarily removed, the surface of the door was revealed as it must have appeared in the early post-medieval period. Beneath the lower lock was the well defined scar and nail-holes relating to a section of the lost iron edge-band. Whether it was removed when the lock was fitted, or lost at an earlier date,

*Figure 98: Rochester Cathedral, staircase door. Detail of the external face of board 4, after removal of the box-lock, revealing traces of the primary polychromy: an area of white ground to the left, specks of vermilion-coloured paint (circled) and reddish-brown paint where there had once been an iron edge-band (oval encirclement). © John Crook*

is uncertain, but directly underlying the band was a strip of reddish-brown paint (Fig. 98). That in turn overlay white paint, which must have been primary and applied while the door was in the carpenter's workshop. This discovery confirms that the outer face of the door was not covered with hide during its manufacture.

Scrutiny of the exterior face generally, revealed extensive traces of the white ground, and in some places it exhibits a distinctly pink tinge, which is doubtless the result of the overlying red paint having leached into the white. Red is only represented by tiny specks that are vermilion in colour, unlike the strip under the edge-band, where the rusting iron had discoloured the paint to a dull reddish-brown.

Three tear-shaped candle-burns are present on board 2 in the lower quadrant of the upper iron circle (Fig. 95B). These apotropaic marks are commonly found on doors, including that in the Westminster vestibule. Also, on board 1, between the upper and middle St Andrew's crosses, is a sizeable patch of rough tool-damage, which appears to result from the use of a chisel. Whether this occurred while the door was being fabricated, or at a later date, is indeterminate. Since there are two adjacent keyholes in board 5, the damage may be connected with a post-medieval lock.

off-course and break through the surface of the timber, disfiguring the door. Hence, the only viable option for measuring the tree-ring growth pattern was by studying the exposed end-grain of each board.

When dealing with a door constructed of vertical planks, end-grain was only exposed at the top and bottom. Since the bottoms of historic doors are invariably decayed and ragged (and often incorporate sections of replaced timber), they are not suitable for dendrochronological analysis. The upper ends of the boards, on the other hand, are generally in much better condition, but still need preparatory work before the ring sequence can be 'read'. Cleaning up the end-grain, to render every growth-ring crisp and clear for accurate measurement, could be done with a sandpaper-block, a chisel or a scalpel. The pioneering dendrochronologists did just that to both the Westminster and Hadstock doors.

However, there were still obstacles to overcome: the Hadstock door has a semicircular head, and the accuracy of mechanically measuring ring-widths on the curved ends of boards is not great. Although the Westminster door is now flat-topped, the ends of the boards are split and battered through age, and thus it was not possible to measure full, uninterrupted ring sequences, except on two boards. The integrity of dendrochronological dating depends upon measuring unbroken sequences of sixty or more growth rings, preferably on several boards. Gaps in the ring sequence are not admissible.

### Trials at Westminster

John Fletcher and Cecil Hewett went to Westminster Abbey in July 1972 to study structural timberwork, and they examined the vestibule door. Hewett opined that both it and the metalwork were early Norman and suggested a date-bracket of c. 1060–80. Fletcher then wrote to Dr Arnold Taylor (Chief Inspector of Ancient Monuments, Ministry of Works), suggesting that the door should be photographed and the visible tree-ring sequence recorded on the tops of the boards.[4]

Fletcher organized the first English dendrochronology symposium in July 1976, part of which took place in Westminster Abbey. In the following year, further reconnaissance visits were made to the Abbey, to assess which artefacts might respond to dendrochronological analysis. The vestibule door had already attracted his attention and he subsequently visited twice, to measure the ring-sequences on the upper ends of board nos 1 and 2. About one hundred rings were recorded, including a substantial amount of softwood on one board.

In 1980, he informed the Historic Buildings and Monuments Commission (predecessor of English Heritage) that he was able to determine the felling date of the tree that yielded the two boards to lie within the bracket 1100–10.[5] This was later than some scholars anticipated, and it raised a serious question. Why, if the door dated to c. 1100, was it so fundamentally different in style and construction from all other English doors of that period? Something was not right, and no further progress was made for a decade.

## Trials at Hadstock

In 1975, I paid periodic visits to Hadstock to continue studying aspects of the church's structural archaeology in association with Hewett, which included recording the timberwork of the roof, window frames and doors. The antiquity of the church and its north door were still contested issues, over which local historians had robust views (p. 88). Consequently, we considered that it was worth attempting to obtain a tree-ring date from the exposed upper ends of the boards of the door, even though they were cut to a semicircular profile. Photographic measurement was not appropriate in this situation.

Back in Oxford, I discussed the proposal with John Fletcher, who readily agreed to attempt to date the door. We assembled at Hadstock in August 1975. At that time, the method of measuring tree-rings on boards involved the use of an optical graticule, which is an

*Figure 99: Hadstock church, north door. Dr John Fletcher using a graticule to measure the tree rings on the arched top of the door, 1975. Author*

eyepiece with an integral scale, used by biologists to measure cell sizes. The task was laborious but two of the board-ends were fully measured (Fig. 99).[6] The operation was not as successful as we had hoped, and Fletcher had difficulty in matching the recorded measurements to any master tree-ring sequence. Eventually, he thought he had found a match with an early 11th-century sequence, and tentatively suggested a felling date for the timber in the 1020s. It was not secure enough to be published, and there the matter rested; he planned to return to Hadstock, recheck the measurements and tackle the other two boards. Sadly, before that could happen, Fletcher died in 1987.

## Return to Westminster

The vestibule door at Westminster and the north door at Hadstock were both contenders for the title of 'the oldest door in Britain'. It was clear that, although they were markedly different in style, they were not far apart in date, and the only way to resolve their relative ages would be by dendrochronology. We have just seen that attempts to record and date the tree-ring sequences of both doors in the 1970s were unsatisfactory because there were problems that could not be overcome with the equipment available at the time. But a technological advance – described below – now offered fresh hope for accurate dating.

As it happened, in 2000 I was engaged on writing a new guidebook for English Heritage on the Chapter House and Pyx Chamber, and obtaining an accurate date for the vestibule door was an obvious goal to aim for.[7] So I drew up a proposal for dendro-dating the door. Although the chapter house and Pyx Chamber are structurally part of the medieval abbey, by a quirk of history they are not the property of the Dean and Chapter of Westminster. They have belonged to the Crown since the reign of Henry III, and at the millennium were independently managed by English Heritage. The ownership of the vestibule was less certain, both parties laying claim to it, and a legal challenge was in progress.[8] Since the door hangs in the south wall of the vestibule, it lay within the disputed area, and consent to carry out the dating was initially approved but then rescinded on the grounds that it might prejudice the legal proceedings. Stalemate obtained for several more years, until counsel advised that neither party was likely to win, through a lack of solid evidence. Eventually, a 'local management agreement' was put in place, whereby English Heritage assigned the day-to-day running of the chapter house and Pyx Chamber to the Dean and Chapter, whilst the Crown retained ownership. That arrangement remains in place today.

The year 2005 was celebrated nationally – and in particular at Westminster Abbey – as the likely millennium of the birth of Edward the Confessor. That signal event presented a suitable justification for resurrecting the proposal to investigate whether the Westminster door was fabricated during his reign, and was therefore a component of the Confessor's new abbey, the construction of which began in the mid-1040s. The scope of the revised dendrochronology project was now broadened to embrace other historic doors in the Abbey. Consequently, a request was made to English Heritage to commission a programme of dendrochronology on six doors, several of which had been the subject of early attempts at dating by Fletcher.[9] He measured the exposed rings on the tops of two boards, which were found to cross-match together but, as noted above, failed to provide a date sufficiently reliable for publication.[10] More evidence was needed and that could only be obtained from measuring the other boards, but their ring-patterns were inaccessible, without cutting into the door, and that was out of the question.

## An improved method of sampling for dendrochronology

A seminally important new development in the process of recovering samples for dendrochronological dating made it possible to capture the essential evidence without causing visible damage to the door. Shortly before the turn of the millennium, a major technical advance in timber coring was made by Fletcher's successor, Dr Daniel Miles of the Oxford Dendrochronology Laboratory. He designed and constructed a drilling rig that could be clamped to a door and carry out micro-boring transversely through its boards.[11] The outside diameter of the hollow drill is 8 mm, extracting cores that are only 5 mm in diameter. To prevent the drill from jamming with sawdust, compressed air is piped into the bore.

If the boards in a door are flat and abut one another squarely, it is possible to continue the coring operation through adjacent timbers. This revolutionized the dendrochronological-dating of doors, chests, panelling and other artefacts composed of relatively thin boards where access to measurable end-grain was not feasible. In the space of barely three years, the three doors for which claims had variously been advanced to be the 'oldest' in England were sampled by Miles.[12]

## Obtaining secure dates for the Rochester, Hadstock and Westminster doors

### Rochester Cathedral, 2002

The stair-turret door in the north-eastern transept has been described in chapter 6 (Figs 93–98). Although the turret is part of the east end of the church, built *c.* 1200–15, the door is clearly older, and is known as 'Gundulf's door'. He was the Norman bishop responsible for the reconstruction of the cathedral priory, 1077–1108, but little of his church survives today.[13] It was generally presumed that if this door was relict from Gundulf's church, it would most likely date from the late 11th century. Since no English church doors were securely datable to that era, it was worth testing by dendrochronology.

Daniel Miles was commissioned in 2002 to carry out dendrochronological dating of the door.[14] The four primary boards were all cored, and three were demonstrated to have originated from a single tree, but the fourth was from a different source. The felling date-range for the timber was established as 1075–1107, and hence the door's association with Gundulf established.[15] It was described by Miles in 2002 as the 'earliest scientifically-dated door found in the British Isles'.

### Hadstock church, 2004

Fletcher's attempt to date the door by dendrochronology in 1975 failed to yield a result that was scientifically acceptable (p. 123), but with the advent of micro-boring, it was now possible to obtain cores that avoided the difficulty of measuring tree-rings around the curved ends of the boards. Arrangements were made by Jane Geddes for Miles to take micro-bores in 2004.[16]

The four boards were all sampled: their ring patterns matched each other, and they were clearly from the same tree. The results confirmed a felling date after 1034, and probably *c.* 1050–75. Taken together, the recorded ring sequence produced a site-master extending from 663 to 1022, making the tree over 450 years old when it was felled. Hence, it started growing at least by 600.[17] The positions of the cores were recorded on an elevation drawing of the door prepared by Adrian Gibson[18] (Fig. 78).

Hadstock now supplanted Rochester as having the oldest scientifically-dated door in Britain.

*Figure 100: Westminster Abbey, vestibule door. Micro-boring for dendrochronology being carried out by Dr Daniel Miles, assisted by Dr Martin Bridge, with Dr Jane Geddes taking notes, 2005. Author, © Dean and Chapter of Westminster*

### Westminster Abbey, 2005

Forty years after Fletcher's pioneering attempt to date the door (p. 122), Miles was able to match his results to reference chronologies with greater accuracy, but 'more material was needed to bolster the t-value matches to acceptable limits'.[19] Consequently, in February 2005, the door was lifted off its hinge-pintles, laid on trestles in the outer vestibule and the micro-borer clamped to it (Fig. 100).[20] Three cores were taken from the upper part of the door, sampling four of the five boards (Fig. 36A). The wide central board (no. 3) was not sampled since it presented only thirty rings because it was one of the outer planks of the tree when it was converted. The boards were shown to have come from the same tree, and hence the individual ring sequences could be combined to create a site-master of 107 rings. Comparison with established British reference chronologies yielded perfect matches, and the rings were found to span the years 924–1030.

The site-master matched exceptionally well with the chronology from Greensted church (Ess.) and some boards from the Tower of London. Since sapwood was present in two of the samples from the door, with the heartwood/sapwood boundary being datable to 1023, a felling date-range of 1032–64 could be calculated. Dendrochronology thus confirmed the vestibule door as being an artefact of the reign of Edward the Confessor, and hence it acquired the accolade of being the oldest scientifically-dated door in Britain.[21]

Thus, we now have three dated church doors belonging to the second half of the 11th century. In broad terms, they are: Westminster, centred on the 1050s; Hadstock, on the 1060s; and Rochester on the 1080s or 1090s. While they all bore decorative ironwork on the exterior face, and share some constructional details in common, they are undoubtedly the products of three different schools of carpentry, despite being from the same geographical region in south-east England. The several types of construction represented by these three doors are discussed in chapter 9.

# 8

# Related hide-covered doors: Copford, Elmstead and Castle Hedingham

Apart from Westminster and Hadstock, other medieval church doors with attached hides were listed and briefly described in chapter 1. Three of them merit illustration and further discussion for their relevance to the Westminster and Hadstock doors: they are at Copford, Elmstead and Castle Hedingham, all in north Essex (Fig. 144). The doors are 12th century and were hide-covered.

## Copford church, Essex (Fig. 101)

St Michael's Church achieved notoriety in the 17th to 19th centuries on account of the supposed 'Dane-skins' attached to its two doors, about which local historians made much ado (p. 130). More importantly, internal restoration work revealed a series of wall-paintings dating from the early 12th century, one of the most remarkable survivals of the Norman era in Essex (Fig. 142).[1] The church was built and decorated *c.* 1125–30, probably as a chapel by the bishops of London, who had long held the manor of Copford. The nave and chancel comprise four undivided bays, originally barrel-vaulted in stone; only the sanctuary apse still retains its vault. Stone vaulting was never a feature of small parish churches in Essex, which elevates Copford to the rank of being truly exceptional.

The westernmost bay of the nave had opposing Norman doorways, and the northern one remains substantially intact. When the south aisle was added in the later 13th century, the side wall of the nave was opened up to form an arcade, and that involved removing the Norman south doorway. It was repositioned in the aisle, where it, and the door, remained until sometime after the restoration of 1877–78, when the doorway was reconstructed and an entirely new door fitted. The Norman door was excessively restored and moved to the north side of the nave (Fig. 102). The

*Figure 101: Copford church, from the south-east, showing the Norman apsidal sanctuary. Author*

*Figure 102: Copford church. North nave doorway, with wire bird-grille closed. © Paul Ravenscroft*

RCHME described it as 'of battens with marks of former ornamental iron-work, 12th century'.[2] Geddes noted that the frame is new but the boards are original and 'densely pitted with nail holes on the front'; regarding the ironwork, she opined 'only the upper hinge may be medieval', and queried whether it was 15th century.[3]

The doors are flat-headed, but semicircular arches abound throughout the church, including in the masonry of the north nave doorway and the small, blocked doorway in the chancel. The inner order of the former was reconstructed in the 19th century, when a limestone lintel was installed and the tympanum above it filled with modern bricks. There appears to be no plausible explanation for this intervention, unless the Norman doors were originally round-headed. Truncation of the south door most likely occurred when the church was extended on that side.[4] Relocating the cut-down door to the north side of the nave could explain why the arch there had to be modified during the Victorian restoration.

### Carpentry

Initially, the nave was fitted with two early 12th-century doors, both hide covered on the exterior and lavishly decorated with ferramenta of unknown design. It has already been noted that there are multiple historical allusions to the 'doors' at Copford, and the following note, to which one of the extant skin samples is attached, clinches the matter. It was written in 1829 by John Cunnington, a lawyer and antiquary from Braintree (Fig. 16).

On the doors of Copford Church, Essex, were formerly preserved several skins said to have been human; they were partly covered by a kind of flourished iron-work which still remains

& appears to have been put on for the purpose of protecting them; the tradition in the parish is that they were the skins of Danes slain in battle in the neighbouring field; the subject is mentioned in the 'Excursions through Essex' published in 1818 and in Morant, v.2 pa: 196, who refers to Newcourt 2: 191.[5]

This is a piece of one of the skins, and was given me by Mr Nathaniel Cobb jnr. who informed me it was taken from one of the doors about 50 years ago by his grandfather, and was the only piece then remaining.

John Cunnington
Braintree 1829

None of the published accounts differentiates between north and south until after the 1878 restoration, when it was reported that the south door had been rehung on the north. Moreover, that is at variance with a description of 1882, when members of the Essex Archaeological Society visited Copford, and Dr Henry Laver, a notable surgeon and antiquary, gave a brief account of the 'Dane-skin'.[6]

The south door is ancient with some remains of fine scrolled iron-work. It was upon this door that there had long been a tradition that the skin of a Dane, who had been flayed alive in punishment of sacrilege, had been nailed, of which some fragments were alleged to have been found beneath the iron-work. Leaving out the question of the tradition that it had been the skin of a Dane, the truth was brought to the test by the late eminent antiquary Mr Albert Way, F.S.A. who ... demonstrated, not only with respect to Copford, but also to Hadstock in this county, and a door at Worcester Cathedral, from all of which small portions of skin had been obtained, which, having been submitted to microscopical examination ... were conclusively determined in each case to be human; leaving no doubt that this dreadful punishment was occasionally inflicted. None of the skin now remains upon the door, but a fragment taken thence many years before, had been preserved, and the present rector, the Rev. B. Ruck Keene, has also succeeded in obtaining another piece.[7]

From the above, we conclude that the Norman south door may not have been relocated until after 1882. Writing in 1898, Tyack also records: 'No skin now remains upon the Copford door, the last vestiges having been removed about the year 1843 or 1844.'[8] There is no mention of what happened to the previous north door, which must also have been Norman. Lacking the protection of a porch, it was probably in poor condition, with little or no visibly surviving hide. Anyway, it was discarded. The south door had been resited when the 13th-century aisle was added, and was protected by a deep porch from the 14th century onwards. Consequently, its preservation would have been better, and this was presumably the door to which most of the historical writers alluded. Likewise, it must have been the principal source from which multiple samples of skin were taken by antiquaries and souvenir hunters.

Today, the resited door is again exposed to the elements (Fig. 103). Some holes and the joints between the boards have been filled with lime putty, and the whole door thickly limewashed. While this has afforded protection, it has obscured superficial archaeological evidence. Nevertheless, it is discernible that much of the

*Figure 103: Copford church. North nave door, exterior (open). Author*

surface is peppered with nail-holes and the ends of small, broken nails. The door is composed of four oak boards, 40 mm thick, with their edges butt-jointed, rather than rebated.[9] There are presumably two or three edge-dowels in each joint, but they are not currently visible. The door had three or more ledges attached to the rear by pegs, seemingly two per board. The ledges have long gone, but the ends of two lines of pegs are visible in the boards.[10] The door is now backed with Victorian diagonal boarding, the same as on the replacement south door. Modern framing strips have also been added to the two upright edges.

### Hide covering and paint

Ironically, Copford has yielded not only the best samples of skin for study, but also of the original polychromy that decorated the exteriors of doors. Five samples are currently known to exist: three are held in the parish church, and two in Colchester Museum.

(i)  Trapezoidal fragment, crumpled and very dirty. Maximum dimension *c.* 75 mm. Attached with cloth tape to a sheet of paper carrying the above description by John Cunnington, 1829, written in sepia ink (Fig. 16).[11] Copford church.

(ii) Cream-coloured fragment, showing the flesh face, measuring 80 × 50 mm (Fig. 104B). The straight edge results from a knife cut adjacent to a hinge or other fitting.[12] Rust stains around the edges relate to positions where the hide was pierced by nails and clench-bolts. Copford church.

(iii) Roughly rectangular fragment with irregular edges, measuring 70 × 40 mm (Figs 104A and 104C). Clear knife-cuts along the top and right-hand edges. The colour of the underside varies from pale grey to buff, with pink tinges in places. On the upper face a ground of gesso or white lead paint is present over most of the

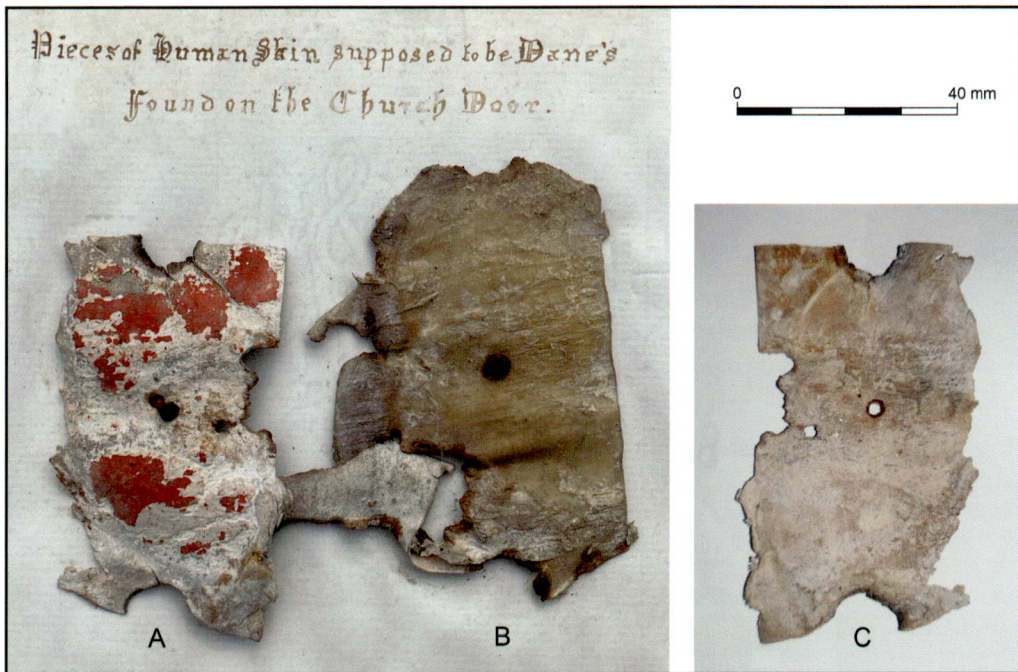

*Figure 104: Copford church. A, hide fragment (iii), with part-holes for two clench-bolts and nail-holes. Red paint with a white undercoat survives on the outer face. B, hide fragment (ii), showing the flesh side, with one cleanly cut edge and part-holes for two clench-bolts. C, hide fragment (iii), flesh face, with potential traces of the glue that adhered it to the timber. Author, courtesy of St Michael's Church*

fragment, and that in turn carries a layer of bright red paint, being the intended decorative finish on the door. There are some rust-marks at the edges, including parts of two circular holes through which clench-bolts passed. Copford church.

(iv) A squarish fragment, measuring 37 × 37 mm (Fig. 105). The lower edge has been cut with a knife, probably alongside an iron strap. The outer surface is almost entirely covered with gesso or white lead paint, applied as a ground for the red finish that survives in part. The underside of the hide is stained and bears the impression of wood-grain from an oak board. Colchester Museum.[13]

(v) Fragment: not seen. Colchester Museum.

## Ironwork

The present north door carries two large, florid hinges of virtually identical design; the upper one is medieval, but heavily rusted, and the lower is Victorian (Fig. 103). Their design and date are problematic: they are not typical of the 12th or 13th century, but nevertheless bear a passable resemblance to a hinge illustrated on a Norman door in a Cotton manuscript of *c.* 1150.[14] They are reminiscent of 'cut-out' designs, particularly a pair of hinges at Abbey Dore (Heref.), which Geddes dates to the mid-14th century.[15]

*Figure 105: Copford church. Hide fragment (iv), taken from the south door. A, decorated outer face. B, inner (flesh) face; the dark band is staining caused by contact with an oak board. Key: (x), part of a perforation where a clench-bolt passed through the hide, leaving a characteristic rust-stained edge; (y), a small nail-hole, also rust stained. Images, Douglas Atfield, © Colchester Museums*

It is difficult to assign these hinges a convincing place in the architectural history of Copford church, and they could be alien. Neither performs a function today, since the door is hung on 19th-century hinges attached to its rear face.

The multiplicity of nail-holes in the boards confirms that the door was heavily adorned with decorative ironwork, but nothing is known of its design. The majority of the decorative motifs appear to have occupied the middle and upper regions, and the many small nail-holes within and around the scrollwork of the upper hinge must surely be relict from the primary scheme of decoration, before the present medieval hinge was fitted.

## Elmstead church, Essex (Fig. 106)

St Anne and St Laurence's Church is situated a short distance east of Colchester. It is of modest size, exhibits architectural features from the Norman period to the 15th century and has opposing entrances close to the west end of the nave. The Norman north doorway is of simple form, having a semicircular head partly built with salvaged Roman tiles (Fig. 107A). A date of *c.* 1100 has been suggested by previous writers, but the presence of an outer ring of rubble-stone voussoirs could indicate construction in the 11th century.

*Figure 106: Elmstead church, from the south. Author*

The doorway had long been out of use and was externally blocked, until it was re-opened in 1935. The rendering was stripped, the blocking removed and both faces of the ancient door revealed. It had two large strap-hinges with attached C-scrolls and an assemblage of other decorative ironwork. The lower half of the door was in very poor condition and two-thirds of the hinge was missing, as were many of the former *appliqué* decorations. Some were attested by scarring on the outer face of the boards.

The vicar in 1935 decided not to attempt to repair it, but to have a replacement door made and hung in the opening. He not only caused the hinges to be replicated, but instigated an archaeological reconstruction of the complete pattern of decorative ironwork, based on residual markings on the timber. The Norman door was conveyed to a joiner's workshop, to inform the construction of the replica.[16] An account of the discovery and restoration was published by an antiquary, Montague Benton, who reported that the joiner 'extracted what were undoubtedly fragments of skin of some kind from beneath the ironwork; these were put aside for expert examination', but were almost immediately lost.[17]

### Carpentry

The Norman door measures 1.05 m in width, and in its truncated form is currently 1.92 m high, but would originally have been *c.* 2.30 m; it consists of four vertical oak boards, *c.* 28.5 cm wide by 35 mm in thickness (Fig. 108); the rebates are 25 mm wide.[18] Some

*Figure 107: Elmstead church, north doorway, 2024. Exterior and interior views, with replacement door of 1935. Author*

shrinkage has occurred and gaps have opened between the rebates. The boards are not edge-pegged to one another, but counter-rebated. This is an uncommon variant of rebating that involved reversing the direction of the lap at intervals between adjacent boards. Thus, the rebate on the left-hand board may initially lap over that on the right-hand board, for part of the height of the door, and then change direction, so that right overlaps left. This process of reversal may occur twice, or more times, on each abutment between boards. The point where the reversal occurs is marked on both faces of the door by a 'dog-leg' in the joint, known as a joggle. The Elmstead door has only two reversals: one set of joggles occurs slightly above the springing-level of the arch, and the other was close to the bottom of the door (Figs 108 and 112).[19] The relatively modest size of this door did not require further counter-rebating between

*Figure 108: Elmstead church. Exterior of Norman north door, now displayed in the church. Author*

these points. The purpose of counter-rebating was to lock adjacent boards together, so that they could not slip downwards, thereby allowing the door to distort under its own weight (*i.e.* to 'drop'; p. 151).

The boards exhibit fairly tight grain, are largely knot-free, and do not retain any sapwood on their edges: they are potentially suitable for dating by dendrochronology. The well preserved upper end of board 2 displays a series of pronounced transverse lines, being the kerf-marks of the saw that cut the boards (Fig. 109). Normally, these lines would have been eradicated by finishing the surface with a plane or shave, but this particular area escaped such treatment. Similar evidence has been recorded on the Westminster vestibule door (p. 70).

The boards were originally secured together by two narrow, recessed oak ledges on the rear face. Ledges were often not synchronized with the positions of the hinge straps, but in the case of the upper one at Elmstead synchronicity was effected

*Figure 109: Elmstead church, north door. Detail of the upper end of board 2, showing horizontal kerf-marks left by the saw during the timber conversion process. Note also the joggled joint between counter-rebated boards 2 and 3 (lower right). Author*

*Figure 110: Elmstead church, north door. Rear of the arched head, showing joggles in the counter-rebated boards and the tapered, dovetail ledge that binds them together. The brown paint is 18th century. Author*

(Fig. 110). The nails inserted into holes in the iron hinge-straps were driven through the boards and into the ledge, but did not protrude through it. Consequently, the ends of the nails were not clenched or fitted with roves. The ledge is D-shaped in cross-section and is partly housed in a shallow trench cut into the back of the door (Figs 110 and 111). The trench is dovetail-shaped in cross-section, so that once the ledge had been driven into the channel it could not slip out. Moreover, the ledge and its housing are not parallel-sided: they taper from 60 mm to 40 mm, east-to-west. These two features ensured that all the boards were firmly locked together (Fig. 112). This is a variation on the unobtrusive flush ledges of the Westminster door, but achieved the same end (p. 75).

Although the lower ledge and its housing have long been lost, they were doubtless identical to the upper one, and again located directly behind the hinge-strap. D-shaped ledges occur on the backs of other 12th- and 13th-century doors, but they are normally surface mounted and secured with nails and roves, as seen at Hadstock. The dovetail-trenched ledges at Elmstead are rarer (p. 137).

The door was strengthened just below its mid-point by the addition of a third ledge, which was surface-mounted on the rear face. The site of the ledge is still marked by a series of four drilled holes, 25 mm in diameter, for pegs that secured it to the boards; not being trenched into the rear of the door, this ledge relied on face-pegging alone to hold it in place. The holes are unequally spaced and only occur in three boards: one in the first board, with the stump of the peg still *in situ*; one in the second board;

*Figure 111: Elmstead church, north door. Detail of the dovetail-shaped upper ledge and its matrix, showing in the edge of board 4. The timber has been ravaged by deathwatch beetle. Author*

*Figure 112: Elmstead church, north door. Diagram illustrating the counter-rebating of the boards and fitting of the upper dovetail ledge. Hewett 1988, fig. 3*

two in the third; and none in the fourth.[20] The door was hinged on the west jamb of the opening in the nave wall and formerly secured by a timber draw-bar, the pockets for which remain in both reveals of the doorway. No evidence for a latching device or lock is present on the old door.

When the north entrance became disused, the door itself fell into decay. The lower ends of the boards rotted and they were shortened by about 45 cm. Most of the decorative ironwork below the level of the upper hinge has been lost, including the central strap that stiffened the boards. Midway up the back of the door, on board 2, is a small group of candle-burns; these are apotropaic marks. Remarkably, a small patch of red and white paint has survived on the decayed back of board 4 (Fig. 113), and another patch on board 1, where it is overlain by post-medieval brown paint. These appear to be the only survivors of an early scheme of decoration on the interior of the door. The appearance is strikingly reminiscent of the white gesso and vermilion-coloured paint noted on the Rochester door and on the fragments of hide from Copford (p. 133).

### Hide covering and paint

The exterior of the door was covered with hide, fragments of which were retrieved from under the ferramenta, most likely the upper hinge-strap. The pieces must have been small (less than 3 cm across). In 2024, a careful examination was made of the

*Figure 113: Elmstead church, north door. Patches of white paint on the decayed back of the door, one with the red finishing coat partly surviving. Author*

upper hinge, in the hope of locating further pieces of hide, trapped between it and the boards. In places, there are gaps, some loosely filled with dust and particles of lime mortar; elsewhere, there are platelets of rust, resulting from the delamination of the ironwork. It is very likely that all visible traces of hide were extracted in 1935, but it is not impossible that further small fragments lie undetected beyond the detritus that blocks visual access.

The purpose of the edge-band of iron, found on many medieval doors, was presumably to frame the decorative ironwork (*cf.* Rochester: Fig. 96), but in the case of doors with a hide covering it would also have held the hide firmly in place, preventing it from curling at the edges when the bonding of the glue began to fail. Nothing survives of the band at Elmstead, but its former presence is demonstrated by the continuous line of nails around the edges of the door. Externally, medieval paint would have been on the hide, not the timber; internally, the presence of paint directly on the boards has been noted (Fig. 113).

### Ironwork
The principal iron fittings were the two strap-hinges and a medial strap that served to stiffen the central part of the door. The remaining ferramenta comprised small ornamental motifs. Hewett published a partial reconstruction of the door.[21]

The ironwork was succinctly described by Geddes:

> Much missing, but ghosts and nail holes are clearly visible. Originally the edging band was all around the door, with a row of unwelded scrolls set underneath it. Two C-and-strap hinges with split-curl terminals. The curls form head lappets on an animal with an almond-shaped

eye. The nails on the Cs and straps project 10 mm. Most of the bottom C is missing. The outer rim of the C has scrolls cut from it.

Crescent shapes are placed back-to-back on the strap. Another strap, across the centre of the door, has a double split-curl terminal and two crescents on either side... Over the whole surface of the door are several crescents and S-scrolls, placed at random.[22]

Nothing survives of the edge-band except scars left by the nails that secured it. The two strap-hinges were presumably identical: they each had integrally attached C-scrolls and double split-curl terminals to the straps. The first pair of curls (cut from the strap itself) is of diminutive form; the second pair, forming the terminal, is much larger, with the curls overlapping an iron crescent (again, not welded). The C has small scrolls (barbs) cut from its outer edge, and its split-curl terminals incorporate the head of an animal with an almond-shaped eye. Only one of these heads survives, and that is fragmentary; there were originally two on each hinge. The hinging end of the strap is lost, but it did not wrap around the edge of the door to form a back-strap, as seen at Hadstock.

The medial strap is represented only by nail-holes and a ghost outline on the boards. The strap was double ended, with split-curls that presumably overlapped crescents, as they do in the hinges. The strap's eastern terminal is well defined on board 1 and more tenuous on board 4. The remaining ironwork comprised individual crescents and S-shapes, applied more-or-less at random. The upper hinge-strap (and perhaps the lower one) was abutted, above and below, by three crescents; scars show that the medial strap was similarly abutted by two crescents. Otherwise, the remaining crescents and the few S-shapes are randomly placed. Their apparent disposition can be seen on the replica door made in 1935 (Fig. 107).

All the ferramenta were nailed to the outer face of the door, there being no clench-bolts, roves or the clenched ends of nails visible on the rear face. It meant that the nails had to be no more than 35 mm in length. Two distinct types of nail are represented: those with large round heads, 5–7 mm in diameter, and those with smaller squarish or pyramidal heads. Geddes commented on

*Figure 114: Elmstead church, north door. Upper hinge and C-scroll, showing the projecting heads of nails that could not be driven to their full depth, on account of the holes in the ironwork being too small. Author*

a curiosity, not noticed on doors elsewhere, namely that some of the nails in the upper hinge-strap and C-scroll were not driven in for their full length, with the consequence that their heads and 8–10 mm of their shanks project (Fig. 114). The explanation for this anomaly is that the tapered shanks of the nails jammed in the holes in the strap and C-scroll.[23]

## Castle Hedingham church, Essex (Fig. 115)

St Nicholas's Church lies close to the Suffolk border, and is one of the county's major Romanesque aisled churches.[24] It is the focus of the medieval market town, lying in the shadow of a Norman castle keep, built by the earls of Oxford. The chancel, with its flat pilaster-buttresses, is the earliest part of the building and probably dates from the mid-12th century. Geddes suggested a date in the 1270s or 1280s for the body of the church.[25]

The church has three primary doors, dating from the second half of the 12th century, and all were once heavily decorated with ironwork. The earliest is in the south wall of the chancel, and the nave has opposing north and south doorways.

*Figure 115: Castle Hedingham church, from the south-east, showing the Norman chancel and its original door. Author*

## Carpentry

The doors display evidence of serious weathering and were crudely repaired in the past, destroying the evidence for the construction and decoration of their lower regions. Rows of large-headed nails were introduced as part of the restoration, and the backs of all three doors are inaccessible to study.[26] The north and south nave doors are each constructed from several boards, joined by counter-rebating with multiple joggles. The longer boards had at least seven joggles on each edge. As Hewett observed, using so many planks – some of them very narrow – would have been unnecessarily time-consuming and costly. The south door is the largest and is composed of eight planks, when it could have been made with five, and the north door with six boards, instead of four (Figs 116 and 117A). Added to this, Hewett reported that the counter-rebates were not of the usual square-edged type, but moulded in cross-section, an unexplained feature that is neither decorative nor functional.[27]

## Hide covering

The south door is expressly recorded as the 'skin door', but since the north door

Figure 116: Castle Hedingham church. South nave door. Acabashi CC BY-SA 4.0

was its pair (albeit of smaller size), it is very likely that this too had a hide covering. The south porch provides protection for the main door, but has no counterpart on the north; hence, any hide covering there would have disappeared centuries ago. The slightly earlier south chancel door does not have counter-rebating, and is exposed to the weather; no evidence remains to indicate if it was ever hide-covered (Fig. 117B). No historic paint is apparent on any of the timbers.[28]

## Ironwork

Although the doors have lost much of their ironwork, they still exhibit details in common and were almost certainly products of the same smithy (Fig. 118). The head-bands of the nave doors share the same design, and the terminals of their hinge-straps are closely similar. The central straps on the chancel and north doors are the same.

*Figure 117: Castle Hedingham church. A, north nave door. B, south chancel door. Author*

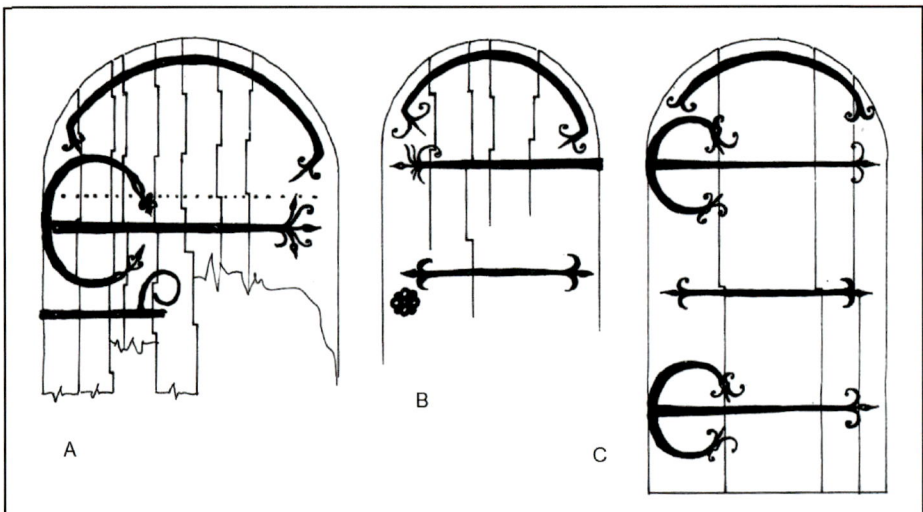

*Figure 118: Castle Hedingham church. Outline drawings of the three Norman doors. A, south nave. B, north nave. C, south chancel. Hewett 1974, fig. 68*

Some of the ironwork has been taken off and reattached; other items have rusted badly, or their fixings have failed, causing them to spring away for the woodwork, leaving gaps. The number of undisturbed locations where small pieces of hide could still be preserved is small. An incomplete scroll in the middle of the south door is one such location, but no sign of a trapped layer of hide is discernible.

# 9

# Early church doors in context: a summary

We have examined the archaeology and structure of the three oldest, scientifically-dated church doors in Britain and described the roles that two of them played over the past sixty years in the study and dating of historic carpentry. The oldest is in Westminster Abbey and was constructed in the 1050s; the second oldest is 45 miles away at Hadstock, and is about ten-to-twenty years later. The third door is in Rochester Cathedral, 30 miles east of Westminster, and that was made c. 1080s/90s. The ancient oak trees from which the doors were constructed grew in south-east England.

All three doors have remained in daily use, in their respective buildings, for between c. 935 and 965 years. They also shared other features, notably being painted red and arrayed with decorative ironwork; the two older doors were formerly covered with hide before being painted. According to oft-repeated local legends, the hides were the skins of human beings who had been flayed as punishment for committing acts of sacrilege. They have usually been referred to as 'Dane-skins', in the belief that the unwilling donors were Vikings.

Despite these shared attributes, the three doors could hardly be more different in construction and appearance, and are products of three separate schools of joinery. This naturally prompts the question: why are they so different, and what is known about 11th-century joinery elsewhere in England? Although many hundreds (perhaps thousands) of churches were built, in stone or timber, during that century, very little of their carpentry and joinery has survived. We have the extant nave of a log-built church at Greensted (Ess.), with a felling date-bracket of 1063–1100, established by dendrochronology.[1] Some structural timbers relating to roofs and floors have survived in a few churches, but they are mainly datable to the early 12th century, as at Kempley.

Extant artefacts of 11th-century joinery are effectively limited to church doors and a few window frames. The doors at Westminster, Hadstock and Rochester have been described in detail, and Geddes considered that a door at Heybridge might also be late 11th century.[2] Several more that share joinery details with Hadstock have long been assigned to the early 12th century on the basis of their architectural context, or decorative ironwork, but they could be older. Some of these are unsuitable for dating by dendrochronology, but others could benefit from it (*e.g.* Elmstead).

Close inspection of the Westminster door has revealed much new information about the tools and techniques employed by joiners in the mid-11th century. Similarly, fresh studies of the surviving fragments of hide that covered some doors have revealed their animal species and traces of the painted decoration that they formerly bore. The archaeological evidence now aligns perfectly with the account of fabricating doors, written by Theophilus in the early 12th century.

## Treatise of Theophilus

We are fortunate in possessing a medieval treatise on various aspects of construction and decoration, compiled by a monk who wrote under the pseudonym Theophilus Presbyter. His true name, dates and place of origin are unrecorded, but attracted much discussion by scholars in the late 19th and 20th centuries. Research by Charles Dodwell has demonstrated that the internal evidence points to the manuscript having been composed between the years 1110–40, and that it was written by a Benedictine monk and metalworker, whom he identified as Roger of Helmarshausen.[3]

In Book 1, Theophilus discusses materials and techniques for painting and decoration. Unfortunately, he does not dwell on woodworking *per se*, but in chapter 17 gives advice on the construction of panels for altars and doors. When the joinery was complete and the basic structure assembled, the face of the timber

> should be smoothed with a planing tool [*i.e.* a shave] which is curved and sharp on the inside and has two handles so that it can be drawn with both hands. Panels, doors and shields are shaved with this until they become completely smooth. Then the panels should be covered with the raw hide of a horse or an ass or a cow, which should have been soaked in water. As soon as the hairs have been scraped off, a little of the water should be wrung out and the hide, while still damp, laid on top of the panels with cheese glue.[4]

In chapter 19, he describes how to whiten hide with glue and gypsum; after concocting the mixture, he gives the instruction to

> spread it over the hide very thinly with a brush. When it is dry, spread a little on more thickly; if necessary, spread on a third coat. When it is completely dry, take the grass called shave-grass, which grows up like a rush and is knobby.[5] You should gather this in the summer and dry it in the sun. With it, rub the white [surface] until it is completely smooth and bright.
>   If you lack hide for covering the panels, they may be covered with medium-weight new cloth with the same glue.[6]

Chapter 20 is headed 'How to redden doors; and linseed oil'. It is devoted to colouring doors that do not have a hide covering. Theophilus gives a formula for making and applying red oil paint:

> If you want to redden doors, get linseed oil which you should make in this way ... spread it with a brush on the doors or panels that you want to redden and dry them in the sun. Then coat them a second time and dry them again. Finally spread on top of it the gluten called varnish, which is made in this way.

Alternative methods for making varnish are described in chapter 21, followed by instructions for gilding and polychrome decoration.[7]

Book 2 is devoted to glasswork and Book 3 to metalwork. In the latter, Theophilus briefly describes a wide range of metalworkers' tools and the materials required for their manufacture. For example, he lists the range of hammer types that are just as familiar today. He refers to the manufacture of steel tools with sharp cutting edges, and describes how to make and case-harden files. In the early 12th century the metalworker's tool-kit and supporting equipment in the workshop was impressively sophisticated. The manufacture of hinges and other decorative ferramenta for church doors would have been run-of-the-mill work for well-equipped artificers.

## The construction of Anglo-Saxon and early Norman doors

Examples of pre-Norman structural carpentry are extremely rare, except in the context of quays, jetties, causeways, etc., that have fortuitously survived by being permanently waterlogged in rivers, lakes and fens. But these are works of engineering in heavy carpentry, not refined joinery. Although few church doors earlier than the mid-12th century have survived in Britain, it must be acknowledged that there may be more early doors that are plain and exhibit no historically diagnostic features or decorative metalwork, their ages awaiting confirmation through dendrochronology.

The joinery displayed in early Norman doors varies considerably, and the reasons for this may be chronological or geographical, although no convincing pattern has yet emerged. Doors of 11th- and early 12th-century date that concern us here mostly exhibit two basic design features: first, the outer face is flush and consists of a series of vertical planks (boards), with their edges abutting. Second, the backs of the boards are exposed on the inner face of the door, and are linked together by several horizontal timbers, termed ledges (or battens), affixed to the boards. Most commonly, the ledges on a door number between three and five.

### *Carpenters' and joiners' tool-kits*

Although Theophilus does not describe woodworkers' tool-kits, there can be no doubt that they were just as sophisticated as metalworkers'. As demonstrated by Massey and Reed in chapter 4, the construction of a church door was a work of skilled joinery, and differing levels of technical virtuosity are exhibited. The

Westminster door is the most impressive. The authors have demonstrated that a substantial tool-kit was required, which included axes, adzes, saws, chisels, gouges, mallet, hammers, awls, shaves, draw-knife, plane, rebate-plane, router, marking-gauge, marking-knife and a square. All of these are attested, either by the tell-tale evidence that they have left in early joinery, or by discoveries of the tools themselves in Anglo-Saxon, Viking and medieval contexts, *e.g.* at York, London, Dublin, Thetford and Ebbsfleet.

Tools of many trades have turned up in hoards of metalwork, as at Hurbuck, Lanchester (Dur.), which included Viking carpenters' and farmers' tools of the 9th to early 10th centuries.[8] Several hoards of iron tools have also been uncovered in England in recent decades (*e.g.* at Flixborough and Nazeing) and Continental hoards have further added to the contents of carpenters' tool-chests (*e.g.* Mästermyr, Gotland).[9] Tools are sometimes found as grave-goods in Anglo-Saxon burials, most notably the fine cabinet-maker's plane from a 6th-century grave at Sarre (Fig. 48B), and Viking-age interments in Norway were accompanied by shaves, which is another tool-type for creating a smooth finish on timber (Fig. 52).[10] These grave-goods indicate that the deceased were woodworkers.

The construction of churches flourished in England throughout the 11th and 12th centuries, and the demand for carpenters and joiners must have been prodigious. The sophistication of the joinery and, often, the added ferramenta, leaves little room for doubt that the construction of doors, screens and panelling was carried out in carpenters' workshops, requiring large and sturdy work-benches. Robust trestles may also have been present, but they would not have provided solid support to the underside of a large door when ledges or ironwork were being nailed to it. Driving nails through two layers of seasoned oak required considerable force, and any deflection or movement of one of the timbers would likely result in bending the nail. If that happened, the damaged nail had to be extracted; hence, the carpenter's kit must have included a tool with a forked end, either a claw-hammer or a jemmy. We have noted that when a hinge was being fixed to the Elmstead door, some of the nails were too thick for the holes in the ironwork, and they became jammed before they could be driven to their full depth. Either no drawing-tool was to hand, or inertia prevailed, and the nail-heads were left projecting (Fig. 114).

### Methods of joining boards together

Techniques for joining vertical boards together and attaching the ledges varied. Simply abutting the squared edges of one board against the next is unsatisfactory (but occurred at Copford and elsewhere): it results in small gaps which will increase in size as the boards shrink. Also, some boards will twist as they season, distorting the face of the door. These problems can be minimized by rebating the edges of the boards so that they interlock: shrinkage will not result in visible gaps and twisting should be reduced. The boards used for the doors at Westminster, Rochester and Hadstock all have continuous rebates. The first two are square-edged, but at Hadstock the rebates are splayed. Some doors have V-shaped or tongue-and-groove joints (*e.g.* Little Hormead, Fig. 126).

Rebates and butt-joints can be rendered more secure by inserting short dowels into holes drilled in the edges of the boards, as is the case at Westminster and Rochester, but not at Hadstock.[11] A novel departure is seen in the west nave door at Kempley, where three pairs of loose ('slip') tenons were fitted instead of dowels, and secured with pegs (Fig. 119).[12] The boards of the Worcester Cathedral hide-covered door were also joined with loose tenons. Another method of strengthening the vertical jointing of rebated boards was to alternate the direction of the lap at intervals, a technique known as counter-rebating. This created an interlocked effect, which resisted the tendency to twist, but more importantly it prevented sagging; *i.e.* the boards all remain in register, and cannot 'drop' (slip downwards) under their own weight.

Of the hide-covered doors discussed here, only Elmstead and two at Castle Hedingham display counter-rebating. The technique was widespread in 12th-century doors, and examples are found at Kempley, Old Woking, Stillingfleet, Worfield, Edstaston and Ely Cathedral.[13] Stillingfleet, at least, had a hide covering on the door (p. 18; Fig. 13). Counter-rebating was an ingenious solution to a common problem, but it involved a lot of work. Straight-through rebates could be cut quickly, using a rebate-plane or, more laboriously, with a chisel and mallet. Some doors with counter-rebating, such as Elmstead, Bristol and Kempley, have only two joggles per board: one near the top and another

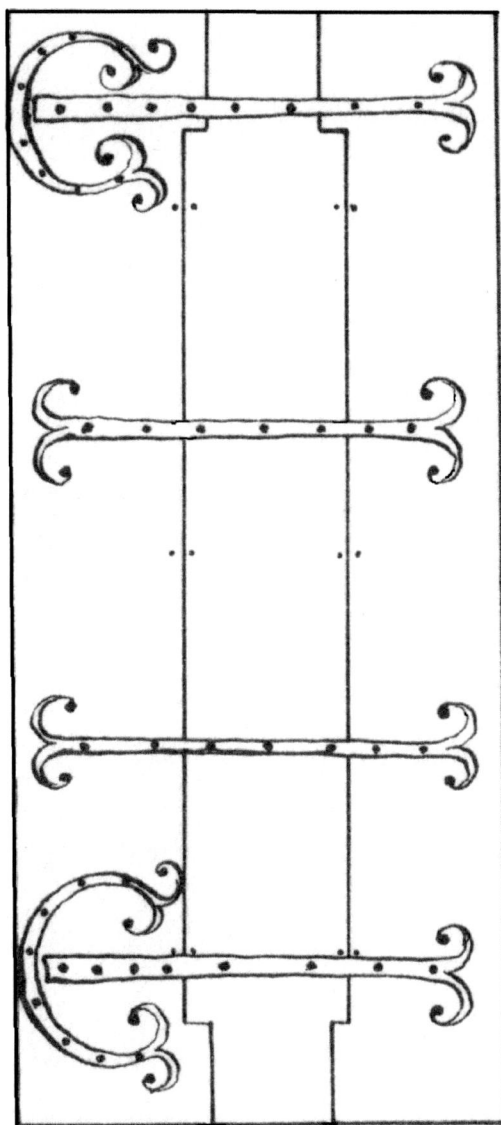

*Figure 119: Kempley church. Exterior of the west nave door, showing the full complement of primary ironwork and the joggled joints close to the top and bottom of the boards, indicative of counter-rebating. Hewett 1980, fig. 41*

close to the bottom (Figs 112 and 119), in which case a plane could be used to rebate most of the uninterrupted section between these extremes. However, doors with multiple counter-rebating, such as Castle Hedingham, Edstaston and Ely Cathedral, would have to be painstakingly worked upon with a chisel and mallet (Figs 116 and 118). The labour involved in counter-rebating doors made from narrow boards was much greater than those with fewer, wide boards. The Bristol Cathedral door comprises no less than eight counter-rebated boards.

A variant is found at Little Hormead, where there is only one line of joggles between boards, and that is at the very top of the door (which has been truncated by cutting off a semicircular head, Figs 123B and 141). This would not have been very effectual and is a curious anomaly.[14] Remarkably, the same feature is present on the only counter-rebated door known in Sweden (see below). Geddes listed thirteen English churches having doors with counter-rebated boards.[15] In date, they span the whole of the 12th century, and the earliest examples are found in the southern counties. Some are plain (*e.g.* Elmstead and Sutton), but have C-hinges with split-curl terminals, and have been assigned to the early part of the century. Although Westminster demonstrates that ironwork of this design was being made from the mid-11th century, the Bristol door, which can hardly be earlier than *c.* 1150–60, has a pair of plain 'Anglo-Saxon' strap-hinges with split-curl terminals and no other elaboration.[16] Whether counter-rebating on church doors began in the late 11th century, can only be determined by dendrochronology.

A single instance of counter-rebating has been recorded in Sweden, in a door from St Olof's Church, Skanör.[17] It differs significantly from the English examples: it is not a work of joinery, but of much rougher carpentry; dendrochronology indicated that the door dates from the second half of the 14th century; and the boards are held together by two wide, flat ledges, each fixed with two rows of clench-bolts with diamond-shaped roves.[18] It is not credible that this complex method of jointing could have been separately invented in two countries, and two centuries apart: more likely the Scandinavian carpenter had visited England in the 14th century, where he examined a door with counter-rebating and decided to copy it himself.

A second Swedish example helps to reinforce this view. A small door in the tower of St Anna's Church, Hörup (Lund diocese), gives a superficial appearance of being counter-rebated, but it is not. Again, it is a work of carpentry, comprising three boards and multiple ledges on the rear. The two joints between the boards are butted – not rebated – and both have a single joggle towards the upper end. Again, the date is likely to be late 14th or 15th century.[19] Counter-rebated jointing does not appear to have been reported in other European countries.

### Types of ledges

A door made of planks needed backing with something that would hold the joined boards tightly together and at the same time prevent the assemblage from distorting. From the early 13th century, this was commonly achieved by attaching a rigid timber framework to the back of the door. The frame might take the form of a regular grid of small squares ('portcullis' framing) or a latticed arrangement; both are seen in the Pyx Chamber doors at Westminster (Fig. 9). Earlier doors were mostly strengthened on the back by fixing horizontal 'ledges' across the boards at intervals, a practice attested in Britain from the Roman period onwards. The form of the ledges, and their methods of fixing, may have chronological significance but the size of the recorded sample is too small to yield credible results.

*(i) Surface-mounted ledges, nailed or pegged*
These are found in a variety of applications: first, the ledges could either be attached horizontally, or diagonally. In the case of the latter, the lower ends should point

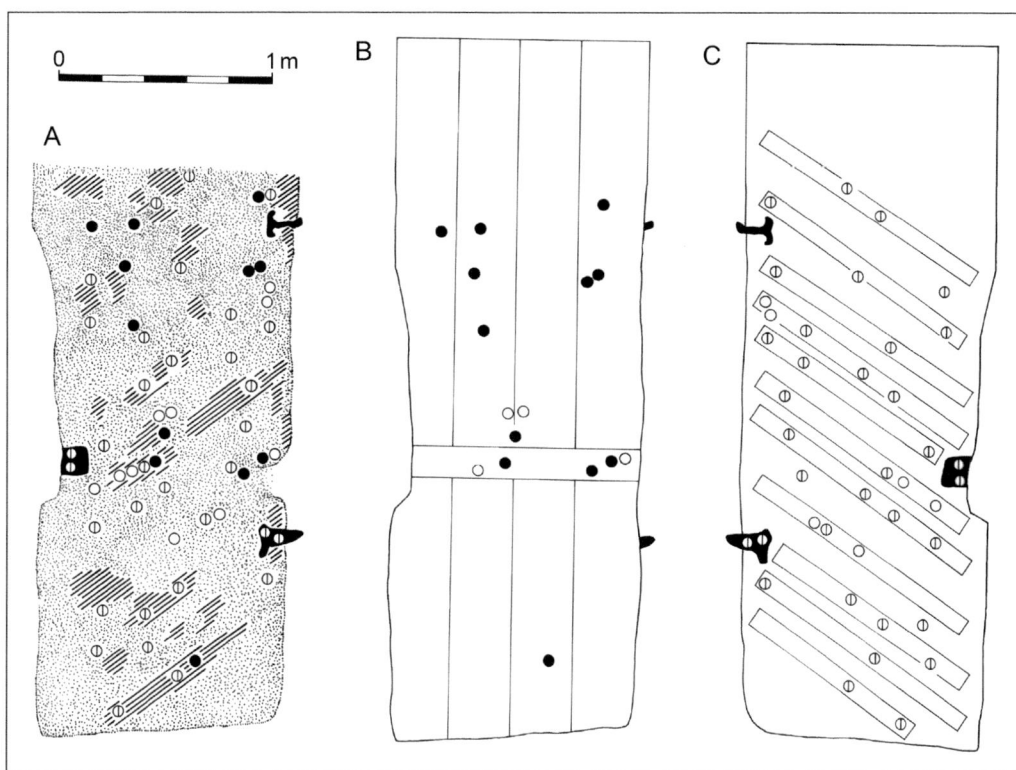

*Figure 120: Pudding Lane excavation, City of London. Interpretation drawings of the vestigial remains of a late Saxon door, comprising vertical boards and multiple diagonal ledges, fixed with nails. A, impression of the door recorded on site as wood-stains (hatched) around a series of nails. Those with their heads upwards are shown as black dots; those with their points uppermost as bisected circles. B, reconstruction of the external face of the door. C, reconstruction of the internal face, showing the positions of diagonal battens and hinges. Horsman 1988, fig. 84*

towards the bottom corner of the door, on its hinging edge. Ledges thus placed not only tie the boards together, but also brace the door against 'dropping'.

The earliest recorded evidence for this type of construction dates from the 1st century AD. A Roman door found in waterlogged conditions at Drapers' Gardens, London, comprised three wide oak boards with four ledges on the rear, secured with clenched nails. It did not hang on iron hinges, but had a stile attached to one edge with timber pivots at the top and bottom (*i.e.* it was 'harr-hung'). Dated by dendrochronology to AD 53–89, the door measured 1.71 × 0.93 m.[20] Vestigial evidence for a door of the later Saxon era was found in an archaeological excavation at Pudding Lane, London, where its 'shadow' was recorded, lying on the floor of a secular timber building dating from the 9th or 10th century. The front comprised vertical boards backed by diagonal ledges, fixed with iron nails; the rusted remains of two hinges and a bolt were also present (Fig. 120).[21] The door was described thus:

> Sufficient remained to allow an almost complete reconstruction. The London door consisted of a single leaf, 2.3 m × 0.8 m × 50 mm, formed from four vertical oak boards *c.* 35 mm thick, secured on their internal faces by eleven diagonal battens 0.72 m × 60 mm × 15 mm. Each of these was anchored to the boards by some four nails. The boards were further secured by a single horizontal wooden batten nailed across the external face of the door.

The excavators commented how the door had racked and the fastening edge had dropped markedly, which would not have occurred if the diagonal ledges had been correctly fitted to act as bracing. The hinges were small and did not have long straps, but were notably similar to the short back-straps at Hadstock, terminating in a split-curl (Fig. 81).

Most early church doors were fitted with horizontal ledges. In many cases there were between three and five, well spaced (*e.g.* Rochester), but others were more numerous and close-set: *e.g.* Chichester Cathedral with ten ledges,[22] Heybridge with seven[23] and Staplehurst with six.[24] The majority of doors of this design had their ledges attached with nails that were driven through the boards and clenched, as at Rochester (p. 115), but some were secured with pegs ('trenails').

*(ii) Surface-mounted ledges with a rounded profile, fixed with nails and clasping roves*
This highly distinctive form of construction is represented by only a few examples in southern England, the finest of which is the north door at Hadstock (Fig. 80). It has four ledges on the rear face, three of which are coincident with the hinges. It also has a 'frame' in the form of a large hoop that runs continuously up both edges of the door and around the semicircular head. The ledges and hoop are all made from timbers, U-shaped in cross-section, and held in place by 170 double-pointed, claw-like roves, through each of which a nail was driven. These clasping roves were used only as decorative washers, and do not perform a structural function.

*Figure 121: (opposite) St Peter-in-the-East, Oxford. Exterior face of the lost west door, showing decorative ironwork. Drawn by J.C. Buckler. British Library, Add. 36433, fol. 665*

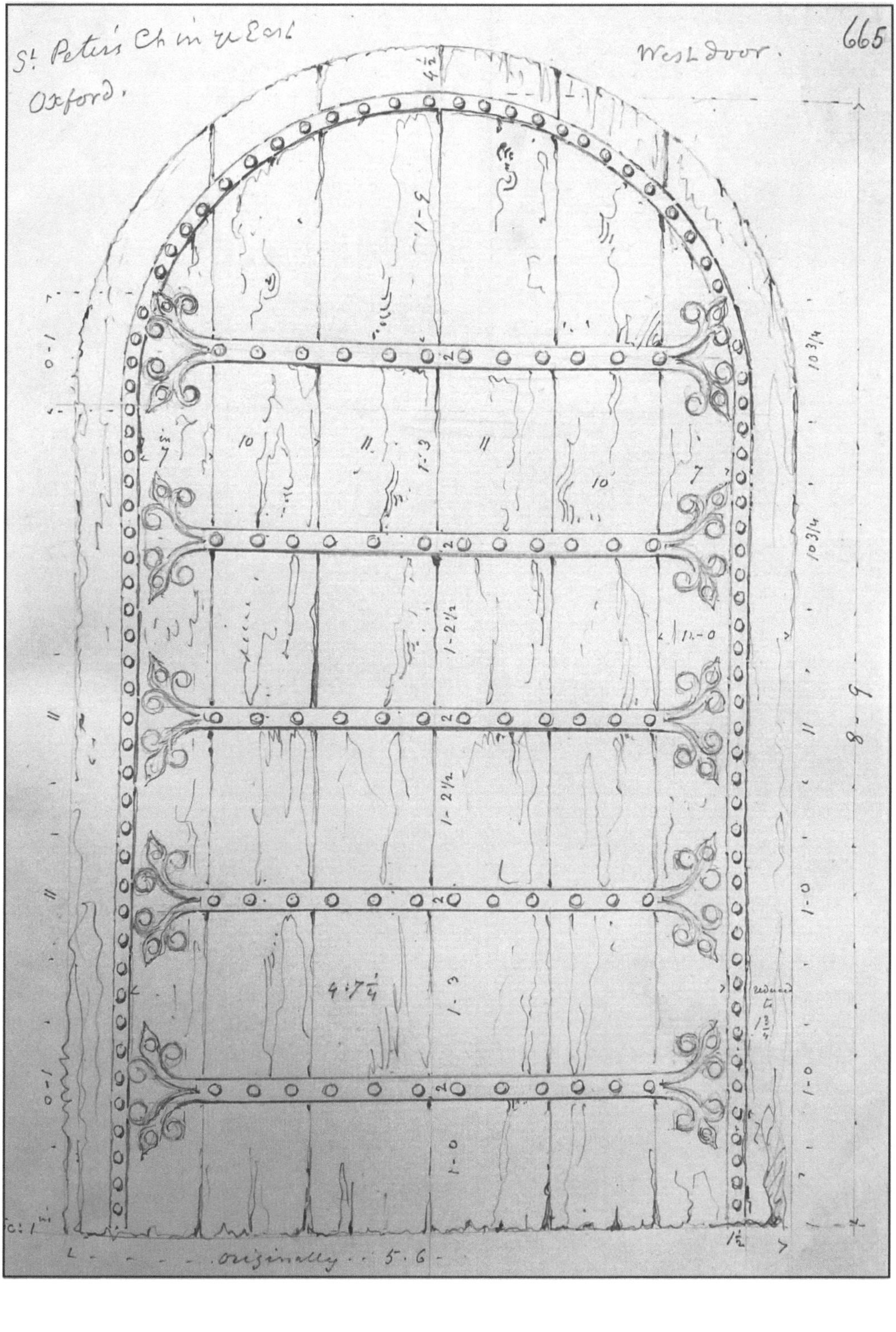
St Peter's Ch in ye East
Oxford.
West door.
665
originally 5.6
8-9

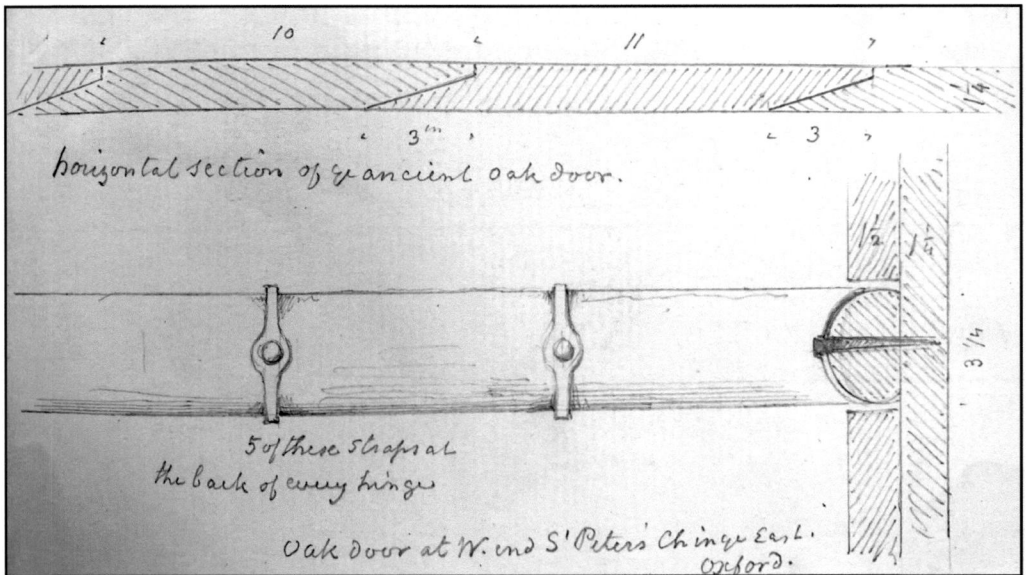

*Figure 122: St Peter-in-the-East, Oxford. Interior face of the lost door, showing details of a ledge with clasping roves; also a section through the splayed rebates of the joints between the boards. Drawn by J.C. Buckler. British Library, Add. 36433, fol. 668*

A closely similar door formerly hung at the west end of the church of St Peter-in-the-East, Oxford, but was regrettably discarded during a 19th-century restoration. Drawings made by J.C. Buckler show ferramenta on the outer face, and details of the fixing of the timber ledges on the reverse (Figs 121 and 122).[25] The door measured 2.67 × 1.68 m, had a semicircular head and was made of six boards, the edges joined with splayed rebates, as at Hadstock.[26] On the exterior was a continuous, flat-iron hoop extending around the perimeter of the door, set well back from the edges of the boards. Unlike Hadstock and Little Hormead, there was no iron inner hoop (Fig. 123), but five horizontal straps, each decorated with split-curls at both ends. They were fixed with a multiplicity of large-headed nails. Buckler's notes imply that these were all hinges, but they are not illustrated as such. The internal face of the door is represented only by a sketch showing details (Fig. 122). It had five semicircular-section ledges, coinciding with the positions of the external iron straps; they were fixed to the boards with nails and exceptionally slender clasping roves. The ensemble was completed with a timber frame at the edges of the door, which presumably included a hoop around the arched head, as at Hadstock.

Thirty miles south of Hadstock lies Buttsbury, where the church has a cut-down and re-backed north door in the nave, composed of five boards joined with splayed rebates; it has three well rounded ledges, two of which are fixed with nails and clasping roves (Fig. 124A). Since it has much in common with Hadstock, Hewett assigned the door to the early 11th century, but Geddes opted for *c.* 1160s, based on the design of the

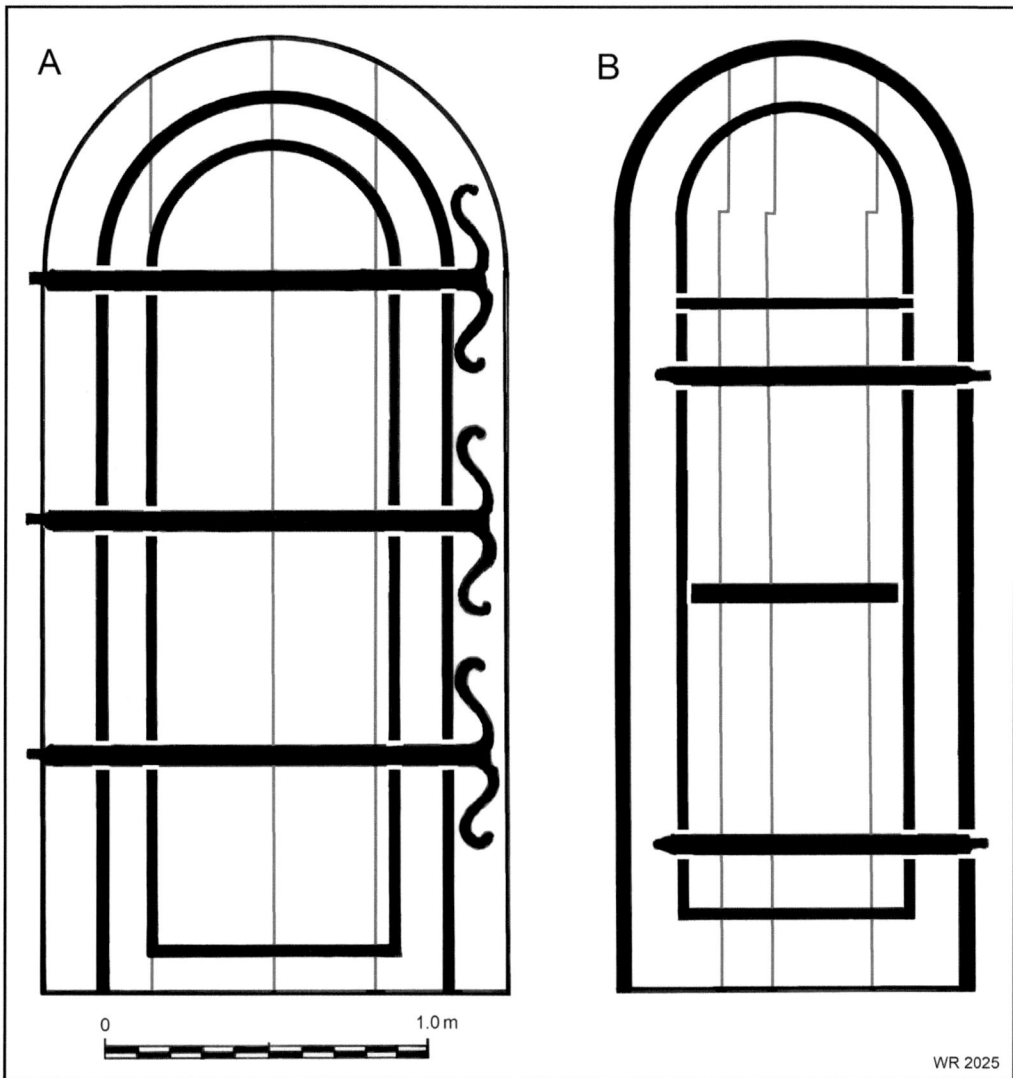

*Figure 123: Church doors fitted with flat-iron banding on the exterior, showing the assembly of the component parts. A, Hadstock, north nave door: reconstruction based on known elements. The west door was almost certainly identical. B, Little Hormead, north door: existing framework, with a reconstruction of the lost semicircular head. Author*

ferramenta on the outer face.[27] When dated by dendrochronology in 2010, the timbers returned a surprising result: the felling date was 'after 1156',[28] which is a century later than its counterpart at Hadstock. Geddes listed eighteen churches that have (or had) doors with rounded ledges, twelve of which were also fitted with clasping roves.[29] Although most of the doors constructed in this manner are in southern England,

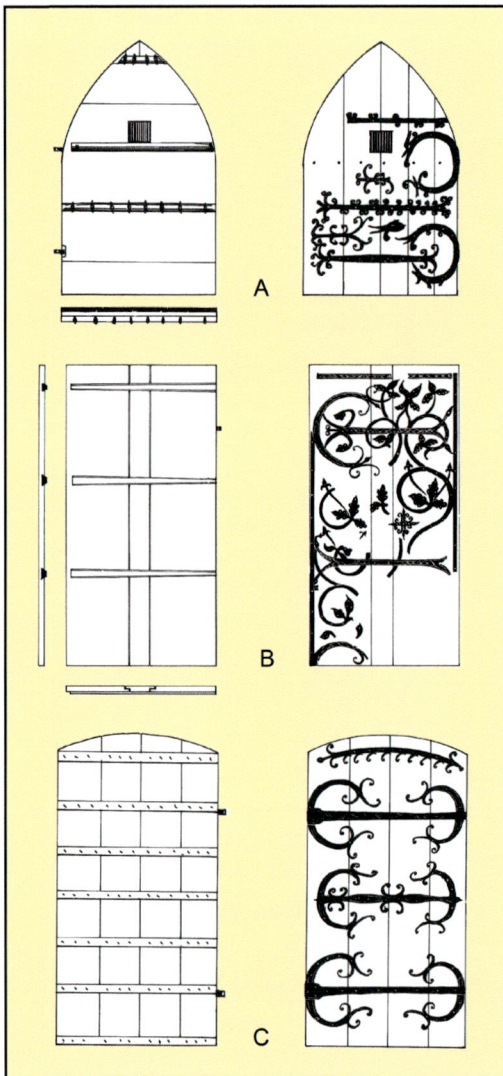

*Figure 124: Back and front elevations of three Norman doors from Essex churches. A, Buttsbury, north nave. B, Eastwood, south nave. C, Heybridge, south nave. Not to scale. After Hewett 1974, figs 66, 70 and 71*

an example occurs in Yorkshire, at Stillingfleet, where the door comprises five counter-rebated boards secured with three ledges (Fig. 13). Apart from Hadstock, Geddes assigned all the doors in this group to the 12th century.

However, attention must be drawn to an important depiction of a door on the Bayeux Tapestry, which was probably made by Anglo-Saxon embroiderers in Kent in the 1070s (Fig. 125). The relevant scene shows William of Normandy arriving outside the gates of Hastings. The town is represented by a compact group of structures, attached to which is an open door. Its size is greatly exaggerated in relation to the buildings and doubtless represents the town's gate. The door has a semicircular head and four transverse attachments: the first and third (from the top) are narrow ledges with their fixings prominently shown, but the second and fourth are distinctly different. They clearly represent metal straps with decorative features, and are best interpreted as hinges, for which they are appropriately positioned on the door. At both ends of the straps are embellishments, the details of which are unclear, except on the right-hand end of strap 4. There, we can discern a split-curl reminiscent of those on the back-straps of the Hadstock hinges (Fig. 81).

*(iii) Wedge-shaped ledges with a rounded profile, trenched into the rear face of the door*

These comprise a minor group, with only six examples currently recorded, at five locations. Three are in Essex, at Eastwood (two doors; Fig. 124B) and Elmstead (Fig. 112); one in Hertfordshire, at Little Hormead; and two pairs of doors at Durham Cathedral.[30] In each case, the ledge has a D-shaped profile and is tapered along its length. The backs of the boards are trenched for nearly half their thickness and are dovetailed in cross-section. The matrix is tapered in plan, corresponding to the taper on the ledges. The rebated boards were presumably edge-pegged (*cf.* Elmstead), laid flat on a bench, and wedged to keep the joints tightly closed while the tapered matrices were cut, using a saw, chisel and router. The ledges would then have been driven into place with a hammer. They needed anchoring, so that they could not work loose over time, and this was achieved in some cases when hinges were fixed on the front of the door, and their nails penetrated the ledges (*e.g.* Little Hormead, Fig. 126).

The same technique was adopted to construct two pairs of massive doors for the nave of Durham Cathedral, 1128–33. Despite their size, each leaf has only three wedged ledges holding it together.[31] The edges of the boards are grooved and jointed with loose tongues. No explanation can be offered as to why the joiner who created the cathedral doors adopted a structural technique that is otherwise only known in a few minor churches in south-east England.

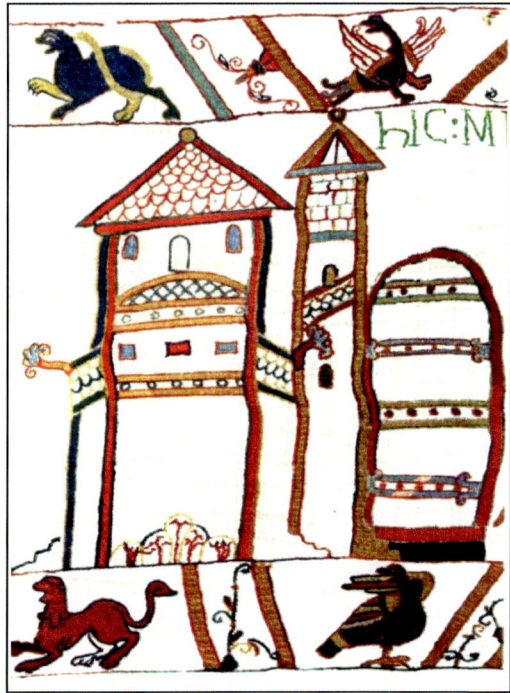

*Figure 125: Bayeux Tapestry. Detail of the scene showing William of Normandy arriving at Hastings. The large door symbolizes the town gate. It depicts a similar form of construction to the north door at Hadstock church. City of Bayeux*

Dovetail wedging was an ingenious invention, requiring a skilled joiner to create the components. The small doors in the Essex and Hertfordshire churches had between two and four ledges. Little Hormead and Elmstead both had only two, but in the latter case there was a third ledge, possibly added later; it was surface mounted and pegged. One door at Eastwood has three dovetail ledges, and the other has four (Fig. 124B).[32]

*(iv) Rectangular ledges trenched into the boards and nailed*
The south door at Heybridge is in a class of its own, having four boards with square, butting edges and seven ledges (Fig. 124C). The latter are also square-edged and trenched into the backs of the boards, but not dovetail-wedged. Instead, the boards are secured to the ledges with multiple nails, driven in from the front and clenched on the back. Geddes dated the door, based on the design of the hinges, to *c.* 1075–1125.[33]

*(v) Ledges fully housed in matrices, flush with the face of the boards*
The Westminster vestibule door is also in a class of its own. As we have seen, the five boards were rebated, edge-pegged and held together by three dovetail-ended ledges that

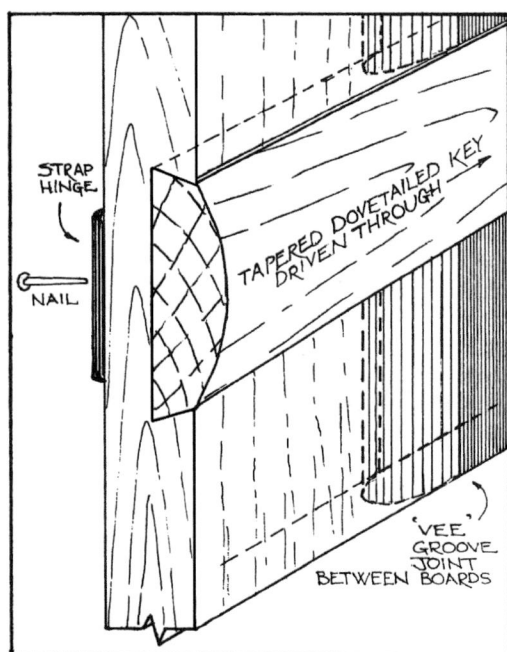

*Figure 126: Little Hormead church, north door. Detail of the fitting of the tapered dovetail ledge, coinciding with the position of a hinge-strap on the exterior face. Adapted from Gibson and Hewett 1983–86, fig. 1*

were fully sunk into matrices cut into the faces of the boards (Figs 23, 25 and 26). Two of the matrices are on the back of the door, and one is at mid-height on the front (pp. 45–50). Being flush on both faces, the door is double-sided without any visual differentiation between front and back. A distinction only manifested itself after the ironwork had been fitted.

No other early medieval door is known to possess these qualities, nor to have been covered with animal hide and painted on both faces. Moreover, it is the oldest church door in Britain, with the date-bracket for felling the timber established dendrochronologically as 1032–64.

### Roves and clench-bolts

In some medieval doors, clench-bolts were employed to hold the ledges and boards together, but in others nails were driven through the timbers and their projecting points clenched. Hinges and other iron fittings were often attached to doors by the latter method. Even if they have subsequently been removed, both roves and clenched nails leave tell-tale evidence on the surface of the timbers. The Westminster and Hadstock doors both exhibit the scars of clenched nails that once fixed metalwork. Although iron fittings were made by blacksmiths, attaching them would have been mostly carried out by the carpenters constructing the doors. Fitting clench-bolts may have been the exception, since that involved a riveting operation and required two men with hammers: one applied pressure to the head of the fully driven nail, while the other cut off excess metal from the projecting point, and then riveted what remained until the rove was securely gripped.

The developing use of roves in door construction has been outlined by Geddes, who identified the visually striking, claw-like roves on the Hadstock door as the earliest examples of their type (Figs 82 and 84).[34] Known as clasping roves, they date from *c.* 1070 onwards, and are found on church doors in many parts of England, but they differ in appearance and function from true roves: clasping roves did not have the function of holding two timbers tightly together, and were not components of clench-bolts. Instead, their 'wings' were wrapped around ledges of bevelled or D-shaped profile, the rove being threaded onto the shank of the nail before it was driven into the ledge. As argued earlier, they were decorative washers (p. 58).

It has hitherto been supposed that the purpose of a clasping rove was to grip the rounded ledge and prevent it from splitting when the nail was driven into it. However, the expanding force of a nail being driven through an oak ledge – without the benefit of a pre-drilled pilot-hole – was unlikely to be contained by the slender 'wings' of the rove. Also it should be remembered that roves were made from thin, flat pieces of iron, and were individually curled to the shape required to fit each ledge. Since the latter were hand-cut strips of oak, finished with a spoke-shave, variations of up to 5 mm commonly occur in their profiles. Therefore, the supposition that roves firmly gripped their respective ledges is untenable. This point is made, *a fortiori*, by the elegant but slim wings of the roves at Hadstock and the even slimmer ones on the lost door from St Peter, Oxford (Figs 121 and 122). Almost certainly, clasping roves were only bent around the ledges, *after* being nailed in place.

As observed by Geddes, the backs of the Hadstock door and others of similar design were intentionally graceful. The neatly formed ledges, steam-bent hoops edging the semicircular tops of doors and the multitude of decoratively shaped roves all contributed to the elegance of the internal faces of these early doors. The original number of roves displayed on the Hadstock door was *c.* 170; on the Oxford door they were more widely spaced and an estimate of at least eighty is likely.[35] Despite the attention paid to its overall appearance, no attempt was made with the Hadstock door to eradicate natural blemishes and tool-marks on the inner face of the boards (Figs 81 and 83).

### Hinges

The typology of hinges was studied by Geddes, who demonstrated convincingly that C-scrolls were a Norman introduction into England, and the Westminster door must represent the very first example, dating from the eve of the Conquest. Hinges and straps with C-scrolls and split-curl terminals have traditionally been assigned to the 12th century, but their presence here from the 1060s is now firmly established.

As Geddes pointed out, not a single example is known of an Anglo-Saxon manuscript depicting C-scroll hinges. Where doors are illustrated, they all have straightforward strap-hinges, often with split-curl terminals. In a depiction of Noah's ark in Ælfric's *Hexateuch* (second quarter of the 11th century), two doorways contain pairs of such hinges in their simplest form.[36] The same manuscript also shows the elevation of a building with a tall, semicircular-headed door with three strap-hinges, each displaying three pairs of lateral curls, spaced along its length.[37] Besides these simple representations are some exhibiting considerable sophistication, a notable example being the Lanalet Pontifical, which depicts the consecration of a great church – plausibly Wells Cathedral – in the first half of the 11th century. The door has two strap-hinges, each with two pairs of opposing scrolls; the central area is filled with a complex *fleur-de-lys* design (Fig. 127).[38]

*Figure 127: Lanalet Pontifical, Rouen. Detail from the consecration of a great church, potentially Wells Cathedral; the officiating bishop is knocking at the door with his pastoral staff. The door is decorated with pairs of facing scrolls on the hinges and other motifs. Bibliothèque Municipale de Rouen*

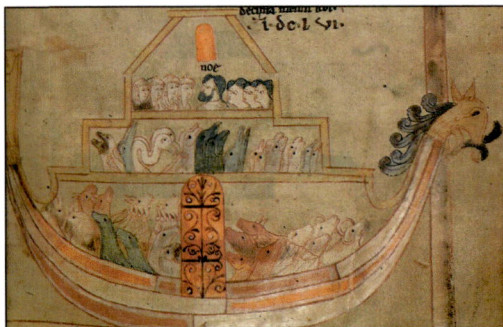

*Figure 128: Ælfric's Hexateuch. Detail of Noah's ark, showing two semicircular-headed doorways, the lower one heavily decorated with ironwork. British Library, Cotton Claudius B.IV, fol. 14r*

A second depiction of Noah's ark in the *Hexateuch*, shows the main door more elaborately decorated (Fig. 128).[39] The two strap-hinges terminate in simple split-curls, but the rest of the door's outer face is crowded with scrollwork, branching from a central stem. At Hadstock, the dense pattern of nail-holes in curving formation could be consistent with similar decoration. Another version of Noah's ark is found in the Caedmon manuscript, dating from *c.* 1000. Here, the entrance door stands open, and people are being beckoned to enter. Only the back-straps of the two hinges are on view, and these are decorated with split-curls (Fig. 129).[40]

*The Last Judgement* from the *Liber Vitae* of Winchester New Minster, *c.* 1020–30, depicts the gates to Heaven and Hell as normal doors. The former is open and it carries two hinges and a central strap, each with two pairs of attached scrolls. These should have been on the outside of the door, but the artist has transposed them to the visible inner face. The closed entrance to Hell is over-compressed in width and has four straps, each with two pairs of scrolls, separated at the centre of the door by a circular motif, possibly an ornate locking device (Fig. 130).[41]

Since pairs of scrolls flanking strap-hinges appear to be a *leitmotiv* of Anglo-Saxon door furniture, an enquiry into its origin would be worthwhile, but that cannot be pursued here. Similarly, the origin of double-ended straps (not hinges) terminating in split-curls or paired scrolls could repay investigation. They appear in Continental manuscript illustrations of the 11th and 12th centuries, and on English doors.[42] Some straps are even found on the backs of doors, as seen in the *Third Life of St Amand, c.* 1169–71.[43]

The interior view of an open tower door in the *Andria* of Terence, originating from either St Albans or Winchester, shows the back-straps of two hinges, each terminating in a small C-scroll. The date of the manuscript is *c.* 1145–55, when C-scrolls were commonplace on the exteriors of doors (Fig. 131).[44]

An exact version of the scrolled hinges shown in the Lanalet Pontifical occurs at Steyning (W. Suss.), where the church is an Anglo-Saxon foundation. The door, which is in poor condition, consists of nine narrow, V-jointed boards initially held together by ledges, but later strengthened with a portcullis frame. Only the upper hinge survives and is potentially original to the door (rather than reused from an earlier door). The outstanding Romanesque

Figure 129: The Caedmon manuscript, probably created in Canterbury. Detail of part of Noah's ark, showing the back-straps of the hinges on the entrance door. Bodleian Library, Oxford, MS Junius 11, p. 66

architecture of the church seems to have influenced historians to date the door to the mid-12th century, but it surely has an 11th-century Anglo-Saxon hinge? If so, it is the oldest decorated church-door hinge recognized in Britain, and John Carter drew it in 1807.[45] This could be another mid-11th-century door, but scientific dating of the boards would be required before making such a claim.

## Early door construction in southern and eastern England: a synthesis

Architectural, archaeological and pictorial evidence all points to the same conclusion, namely that English church doors in the 11th century – and probably much earlier too – were constructed with planks held together by battens attached to the rear face, using clenched nails or clench-bolts. These fixings were very familiar to Anglo-Saxon carpenters, since they had been fundamental components of boat-building for centuries. As might be expected, carpenters of the period produced both vernacular and superior versions of church doors. Only two of the latter survived substantially intact into modern times, but one is no longer in existence (St Peter, Oxford; Fig. 121). Study of the Hadstock door has revealed a previously unexpected level of sophistication in its design and construction.

*Figure 130: The Last Judgement, from the Liber Vitae of the New Minster, Winchester. Details showing the gates of Heaven (A) and Hell (B). British Library, Stowe 944, fol. 7*

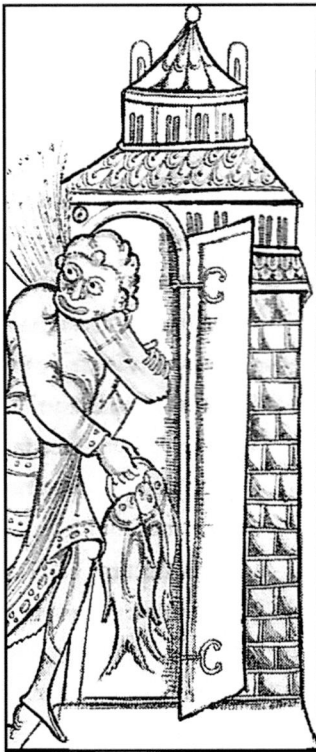

*Figure 131: Scene from the Andria of Terence, showing the back-straps on an open door. Bodleian Library, Oxford, MS Auct. F.2.13, fol. 3*

Anglo-Saxon illuminated manuscripts dating from the first half of the 11th century provide us with many representations of doors bearing decorative ironwork, some of which is impressively intricate, as in the Lanalet Pontifical (Fig. 127) and Ælfric's *Hexateuch* (Fig. 128). The basic hinge type was a strap, terminating in split-curls, but more superior versions embodied elaborate scrollwork. Hadstock's six hinges, from two doors, displayed geometrical patterns formed by numerous small holes in the straps, as well as flamboyant terminal scrolls. Both doors were externally framed with iron strapwork which, together with their hinges, delineated four panels. Each panel was filled with decorative ironwork, all now lost (Fig. 132). The embellishments were set on a background of hide that was painted red. Internally, U-shaped timber ledges and the bentwood hoop mirrored the four-panel divisions in metal on the exterior. The basic design of the Oxford door was similar: it comprised six panels on the exterior defined by iron straps and edged by a flat-band hoop, and six panels on the interior delineated by rounded timber ledges and a marginal hoop.[46]

Considerably, more labour was involved in forming rounded and U-shaped ledges than making rectangular or chamfered battens, and this clearly reflected a conscious desire to enhance the visual aspect of the backs of church doors. Yet greater enhancement was achieved by fitting clasping roves, as washers, under the heads of the nails that attached the ledges and hoops. Like the architecture of the church, the doors of Hadstock exhibited great distinction.

As one would expect, the Anglo-Saxon tradition of door construction was not eclipsed by events of 1066, and the Buttsbury door with its overtly vernacular appearance yielded a felling date for the timber of a century later (after 1156). However, from the early 12th century, innovations began to appear, such as joining boards by counter-rebating and devising new ways of attaching ledges, notably trenching wedge-shaped ledges into the rear face of the door. Both were probably English inventions, but were abandoned within little more than a century, when portcullis and lattice framing gained popularity. Although strap-hinges and decorative ironwork were components of doors long before the Conquest, designs of Norman origin quickly became popular, as exemplified by the C-hinge.

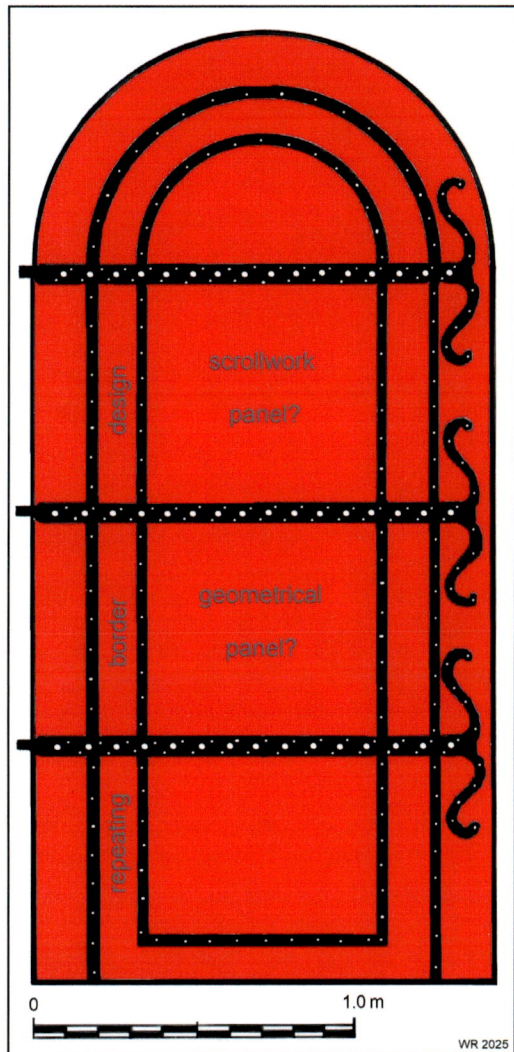

*Figure 132: Hadstock church, north door. Reconstruction of the flat-iron banding, nailed over red-painted hide on the exterior, based on surviving evidence. The four panels and the hoop-band were filled with decorative motifs, seemingly scrollwork and geometrical shapes, all now lost except for one detached fragment (Fig. 139C). Author*

More common in the 13th century were surface-fixed battens, either square or bevelled, like the north transept stair door at Westminster Abbey (Fig. 134). The

*Figure 133: Wells Cathedral, chapter house undercroft. Exterior of the inner door to the treasury, late 1250s. The oak boards were neither covered with hide, nor painted. Author*

Worcester Cathedral door also had bevelled battens and clasping roves, although it is arguably 12th century, rather than 13th (Fig. 3). Clenched nails were the normal method of assembly. The fashion for fitting clasping roves to ledges dwindled in the later 12th century, but there are some notable exceptions, including the chapter house undercroft door and several others at Wells Cathedral.[47] There, the Anglo-Saxon practice of fitting clasping roves over rounded or moulded ledges continued until the latter part of the 13th century, and beyond. In each of those doors the boards are supported by a portcullis frame on the rear, the component timbers of which have moulded arrises and the fixing nails were threaded through clasping roves (Fig. 133).

Finally, we must return to the Westminster vestibule door, which does not fit into any of the categories described: there is nothing comparable in Anglo-Saxon or Norman England that is flush on both faces, or has its ledges totally recessed into the boards. Although Westminster Abbey was being constructed in the 1050s and early 1060s by Edward the Confessor, it was not recognizable as an Anglo-Saxon building. Its design was based on contemporary French Romanesque architecture: in short, it was a Norman building erected in pre-Norman England. The Romanesque design of the abbey church has been much discussed and need not be rehearsed here. Robert Champart, Abbot of Jumièges, was one of the Confessor's advisors, and the king appointed him, first, as Bishop of London, and then Archbishop of Canterbury in 1051. It is not, therefore, surprising that the Romanesque churches of Nôtre-Dame, Jumièges and St Peter, Westminster had much in common.[48]

A large team of carpenters and joiners would have been required permanently on site while the abbey was under construction, and they, like all the other tradesmen, would have been under the direction of Teinfrith, whom Edward described as his 'church-wright'. Although it has often been asserted that Teinfrith was a carpenter, Richard Gem's reassessment of his role has convincingly demonstrated that he was the master mason (effectively, the architect) in charge of the whole construction project, for which the king handsomely rewarded him with an estate at Shepperton (Middx).[49] Teinfrith's ancestry is unrecorded, but his name is probably of Continental Germanic origin.

The vestibule door was made from locally sourced English oak, and the majority of the carpenters employed at Westminster would have been English too, but some French craftsmen were inevitably required to instruct them. It is, however, clear that local carpenters and joiners possessed both the tools and the skills necessary to create high quality utilitarian and artistic works. Consequently, since the Westminster door bears no resemblance to any early English church door, it is reasonable to argue that the design was French. Unfortunately, no 11th-century doors seem to have been preserved in France, to clinch the posited identification.

We can never be certain where the door hung in the Confessor's abbey, but it was most likely in the east cloister range. It was certainly an internal door, connecting two spaces of importance; otherwise it would not have been hide-covered and painted on both faces. The most prestigious location would have been the entrance to the chapter house, from a vestibule. Salvaging, modifying and reusing the door in Henry III's abbey was not a casual or inconsequential act: it represented an instruction, issued probably by the king himself. Hanging it in the new Gothic vestibule ensured that everyone who entered the chapter house had to walk past the door. It stood – and still stands – as a constant reminder of Henry's patron saint, Edward the Confessor.

## Painted decoration on Anglo-Saxon and Norman doors

The decoration of church doors and ironwork received little attention in scholarly works, until Geddes summarized the slender evidence that had been reported in the past.[50] She listed numerous tomb railings displaying historic polychromy and gilding, but was scarcely able to muster a handful of church doors. Since the majority of these hung at the points of entry into buildings from the outside, it is hardly surprising that remains of medieval paint have rarely survived, and any gilding that might once have been on hinges or other fittings has long since been displaced by rust. Furthermore, well-intentioned restoration and redecoration of ironwork on doors, especially in the 19th and 20th centuries, must have destroyed a great deal of vestigial evidence for ancient polychromy. The Victorian dogma that black was the 'correct' colour for ironwork resulted in widespread losses of evidence.

Archaeology has confirmed that Anglo-Saxon wooden shields were covered with hide and painted red,[51] and Theophilus tells us that doors and panels should be covered with animal hide, or cloth, and similarly painted (p. 148). Examples of the techniques described by him have been examined during the course of this study. Vestigial evidence for painting on hides has been confirmed at Westminster and Hadstock, and there are excellently preserved samples from Copford, where the full vigour of the vermilion-like colour was preserved under iron straps that had not rusted

(Fig. 105). Paint analyses have not been attempted on any of the hide samples, but it is considered likely that the base-coat was white lead and the finishing coat was red lead (p. 55). Microscopic examination of the paint on the hide at Westminster revealed a white under-layer and bright red finishing coat, over which traces of a darker red re-painting can be discerned (Fig. 34).

Rochester Cathedral's internal door had no hide covering, yet it displays sizeable areas of eroded whitening and, in a few places, the residue of red overpaint.[52] As a consequence of being in contact with rusty iron, the red has now mostly been stained brown, but occasional specks of the original vermilion-like colour are discernible (Fig. 98). The absence of a hide covering means that joints between boards, tool-marks, knots and blemishes have always been visible, and such imperfections were evidently regarded as acceptable on what was otherwise a high quality door in the heart of the cathedral priory (Fig. 96). Whether Elmstead had paint on the external hide can only be surmised, but there would have been little point in covering the door if it were to remain unpainted. On the back, however, is clear evidence for white coating with red overpaint, applied directly to the boards (Fig. 113).

*Figure 134: Westminster Abbey, north transept staircase. Door of the 1250s not covered with hide, or painted directly on the timber. Author, © Dean and Chapter of Westminster*

Returning to Westminster, exactly the same situation obtained in Henry III's vastly expensive reconstruction of the abbey church in the 1250s. We have alluded to the door with a hide covering that hung at the entrance to the sacristy from the south transept, but was lost in the late 18th century (p. 8). That door was presumably of 13th-century construction and was most likely painted. Of the numerous other doors that must have been made and hung in Henry's new church, only one has survived. It is a small door to a staircase in the north-east corner of the north transept, giving access to the triforium gallery. The external face is of raw oak boards displaying blemishes and shave-marks, but no hide covering or traces of paint. The door has a pair of elegant 13th-century hinges (Fig. 134).

Given that Henry's reconstruction of Westminster Abbey was the most lavish building project of its day, it is surprising that this door (and presumably others) were not treated more sumptuously.

However, archaeological evidence is accruing to reveal that decoration of the fabric was not undertaken at the time of construction. Thus, there is no sign of paint on the ribs or webs of the ceiling vaults, or on the wall diapering that extends over large parts of the eastern arm. Nor is there any evidence for 13th-century polychromy on the walls of the chapter house and its vestibules, or on the vaults and mouldings of the east cloister. These absences stand in marked contrast to the extensive polychromy found in similar locations at Salisbury Cathedral in the second half of the 13th century.[53]

Internal doors in other great churches also fail to display evidence for hide covering or, in most cases, for painting directly on the timber. Being in buildings of high quality, protected from weathering, one would expect more evidence of decorative finishes on Gothic doors to have been preserved. Wells Cathedral provides a contemporary analogue for Westminster.[54] The lower storey of the octagonal chapter house was built in the late 1250s, and the door to its undercroft is made of vertical boards attached to a portcullis frame (Fig. 133). The door is primary but its decorative ironwork appears to be too narrow, and has given rise to the assumption that it was intended for use elsewhere in the cathedral. However, Hewett has pointed out that the misfit may result from an error on the part of the blacksmith, who did not provide long enough straps on the hinges for the ferramenta to be correctly centred in the opening, which has deep returns on both flanks.[55] The door is in excellent preservation, and a careful examination in 2024 failed to reveal any evidence for hide trapped under the hinges, or for paint on the timber.

Credible artistic representations of church doors in England in the 11th to 13th centuries are scarce, especially those displaying colour. A notable exception is the *Douce Apocalypse*, an illuminated manuscript of *c.* 1265–70 that incorporates ninety-seven miniatures.[56] Several of these depict the exteriors of buildings with large doors composed of two leaves: some are shown as plain grey-brown with no distinguishing features, but others have their vertical boards individually defined. One door is helpfully rendered in much greater detail: it comprises a pair of leaves with boldly defined

Figure 135: The Douce Apocalypse, c. 1265–70. Detail from a miniature showing a pair of red-painted doors with elaborate, gilded hinges. Note the prominent joints between the boards, indicating that the doors were not covered with hide before being painted. © Bodleian Library, Oxford, MS Douce 180. CC BY-NC 4.0

*Figure 137: Högby church, Sweden. Detail from the hide-covered door, showing the buckling and losses from the ageing hide. English church doors with hide coverings would have developed a similar appearance over time. Swedish Museum of National Antiquities*

boards, painted bright red.[57] The doors are hung with two pairs of large hinges with ornate, gilded scrollwork (Fig. 135). This image clearly depicts major doors of mid-13th-century date, with paint applied directly to the boards.

The tradition of painting church doors red continued throughout the Middle Ages and into recent times. It is not uncommon to see small areas of dark red paint surviving in the mouldings of Tudor and later doors, but comparable evidence for the 13th to 15th centuries is minimal. The surviving portion of one of the Norman north nave doors at Worcester Cathedral is exceptionally informative on several counts (Figs 2–5). This door had a hide covering only on the interior, which was applied in the carpenter's workshop before the ledges were affixed. Both the hide and the carpentry were painted red, and the extant ledge also exhibits white and black paint, the latter in the form of a diagonally-set

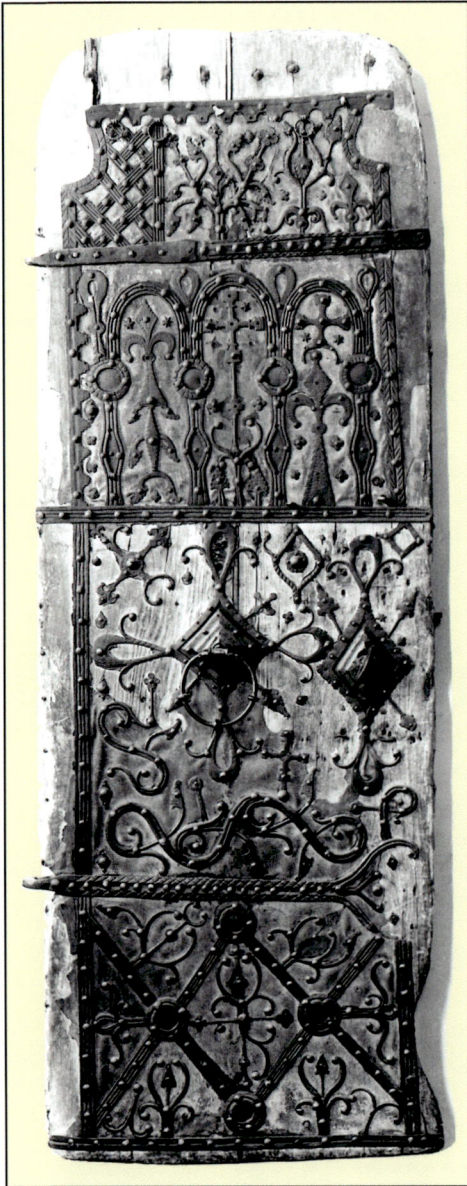

*Figure 136: Högby church, Östergötland, Sweden. Hide-covered door, heavily decorated with ironwork, showing the loss of the hide in the central area, caused by wear, doubtless assisted by petty vandalism and souvenir hunting. Swedish Museum of National Antiquities*

band, which is best interpreted as a remnant of 'barber's pole' decoration. It was claimed in 1894 that this door originated at the west end of the cathedral, and was only later adapted to fit the north porch (p. 4).

The small fragment of hide recovered from behind the sanctuary-ring on the door at Pembridge church is probably datable to the 14th century (p. 18), and Geddes noted traces of red paint under the late 15th-century ring-plate at Raddington (Som.) and on the exterior of a door with decorative ironwork of *c.* 1200–50 at Stanton Long (Salop.).[58]

Although evidence of hide covering and/or paint on church chests is rare, a few examples have been recorded, the most impressive being on the cope chests at York Minster. Chest I has a leather covering on the lid, painted deep red, and trapped beneath the decorative ironwork, but the lid of chest II is painted red, directly on the timber. Chest I was assigned by Geddes to the 14th or 15th century and chest II to *c.* 1275–1300.[59]

Surprisingly little evidence for decorated hides on medieval Continental doors has been published, yet they do exist. They can be seen today in Italy (especially Florence) and Spain, and there are a few examples preserved in museums in Sweden, Switzerland and France, but searching them out is beyond the scope of the present study. The notable door from Högby (Sweden) retains much of its hide covering, a small sample of which was sent to Westminster Abbey in 1967 (Figs 136 and 137).[60] The construction of the door bears no resemblance to English carpentry.

## Hide-covered doors in eastern England: summary and tentative conclusions

The *raison d'être* for covering a shield, door or chest with hide was to enhance the appearance of the object. It concealed the joints between boards, and at the same time hid knots, shakes, tool-marks and other blemishes. Perhaps more importantly, the hide provided a smooth, dense surface, which could be decorated, further enhancing the appearance of the artefact. There is no evidence to indicate that medieval church chests were normally leather covered or painted, and the ironwork was applied directly to the timber. Fixing hides to the exteriors of medieval church doors was clearly selective, and we have no idea what percentage of the doors was enhanced with a covering, or with polychromy. Moreover, antiquarian pilfering in the 18th and 19th centuries, followed by Victorian restoration, resulted in the loss of all visible evidence from church doors that we know were once hide covered.

Work carried out on the Elmstead door in 1935 revealed evidence for a previously unrecorded skin, small fragments of which were discovered behind a hinge-strap. This raises the prospect that there may be similarly preserved fragments awaiting recognition on other doors that are fitted with external strap-hinges and decorative metalwork. The Westminster door provides a case study for the recognition of vestigial traces of hide. Although the skin had been stripped from the visible faces, three pieces of tell-tale evidence remained, demonstrating its former presence. The first is the thin layer of hardened hide that can still be seen with the naked eye between these straps and the timber. The second comprises shallow cut-lines in the face of the boards, alongside the edges of the primary iron strap on the outer face, and beside the secondary hinge-strap on the inner.

In places, it was apparent that the skin had been ripped out from behind the ironwork, leaving a narrow gap between it and the boards. That constitutes a third type of evidence for a lost covering. Nevertheless, the presence of gaps between the ironwork and the boards does not, *per se*, confirm the loss of a hide layer: on most doors, there are areas where the ironwork did not lie perfectly flat on the timber. On the other hand, if parts of the undisturbed ironwork are nailed tightly against the boards, with no gap or intervening layer of hide, then we can reasonably accept this as negative evidence that the door was never hide-covered.

In cases where the door is on the exterior of the church, unprotected from weathering by a porch, the crucial evidence may have been wholly lost, Elmstead being a case in point. The timber is heavily weathered from centuries of exposure to the elements. No cut-lines have survived, and no hide is visible between the ironwork and the boards, but the narrow gap is detectable, although for the most part it is no longer a void. It has become clogged with dirt and platelets of rust arising from the delamination of the iron hinge. Nevertheless, it is possible that further small pieces of hide are still present under the hinge-strap, but obscured from view.

A fundamental question arises: when a door was under construction, who determined whether it should have a layer of hide applied to its exterior before the hinges and any other ironwork were applied? Clearly, fitting and painting a covering of hide introduced an additional cost-element into the door's construction: the patron most likely determined the matter. One might, therefore, assume that such treatment would be found principally in cathedrals, abbeys and other great churches where building-funds were more plentiful than in small rural parish churches. But that was not so.

### Westminster Abbey
The door was covered with cow hide on both sides, and decorative iron hinges and straps were fitted to the exterior alone; there is no evidence for any attachments on the interior face, except a lock, nor for additional decorative motifs on the exterior. We can therefore reconstruct the appearance of the door with confidence, as having four straps, two of which were hinges with C-scrolls and split-curl terminals (Fig. 138). The complement of ferramenta was identical to that on the west door at Kempley church (Fig. 119).

As regards its location, the door was either fully internal, or at least protected from external weathering. The fact that it was not encumbered with framework on the back, but was flat, hide-covered and painted on both faces indicates that it was a communicating door between two spaces of high status. There must have been a powerful reason for the survival of this door and its reuse in the 1250s, when Henry III was funding the construction of the present chapter house and its vestibules. The door was presumably salvaged and set aside for re-use when the Confessor's chapter house and cloister were being demolished. As far as we know, no other furnishing was accorded this treatment, which implies that the door was held in esteem by the king, whose intense devotion to the memory and piety of Edward the Confessor needs no emphasis.[61]

There was seemingly no practical reason why the door could not have been hung, using its original hinges, and not reversed. A possible explanation for taking this extra trouble to re-hang the door was in order to preserve and display in the new vestibule decoration that was on the back face (*i.e.* the side that would have been visible from within the old chapter house), such as a devotional image. While this is speculation, it does not alter the fact that there was something special about the door that caused it to be salvaged and prominently redeployed.

Hide-covered doors were still evidently fashionable in the 1250s, when the sacristy was being constructed adjacent to the south transept (p. 6), but they were not ubiquitous in Henry's new abbey: the stair-turret door in the north transept was not covered (p. 168; Fig. 134).

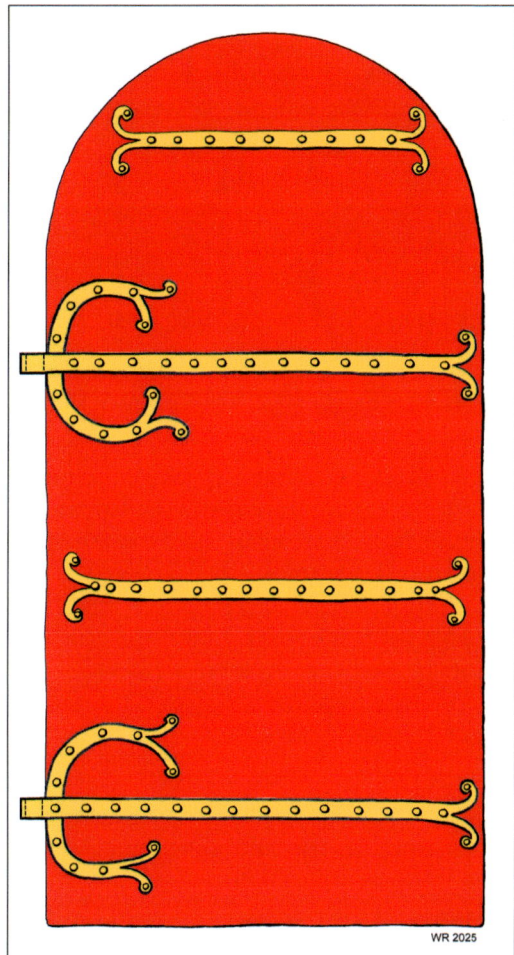

Figure 138: Westminster Abbey, vestibule door. Reconstruction of the original appearance of the outer face. Author, © Dean and Chapter of Westminster

### Hadstock church

Archaeology has demonstrated that this five-celled cruciform church has a complex structural history that is difficult to date. There is general agreement amongst scholars that the decorated masonry of the Romanesque crossing arches and north nave doorway is stylistically datable to *c.* 1080, and cannot be linked to the minster church that Cnut is recorded as having erected in 1020, in thanksgiving for his

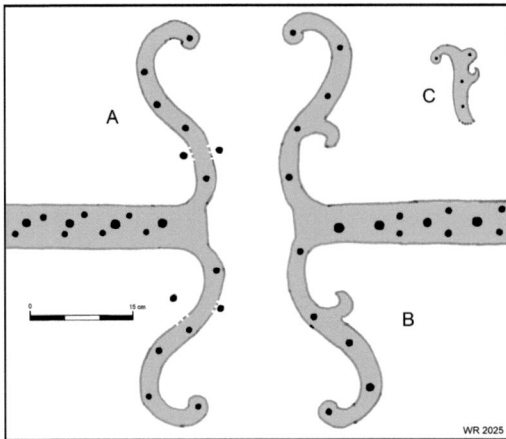

*Figure 139: Hadstock church, decorative ironwork. A, north door: reconstruction of the terminal of the upper hinge, the S-scrolls based on the pattern of nail-holes in board 4 (Fig. 85). Note the two pairs of extra nails flanking the scrolls, relating to lost adjuncts. B, west door: reconstruction of the middle hinge, based on the surviving lower S-scroll. C, north door: fragment of scrollwork in Saffron Walden Museum (Fig. 69). Author*

victory over Edmund Ironside at the battle of *Assandun* (Ashdon) in 1016. Hadstock was already in the hands of Ely Abbey before the Norman Conquest, and I have previously argued that there was a church here in the late Saxon period, the foundations and some of the upstanding walls of which are extant. The possibility that they represent Cnut's memorial church is not proven.

If Hadstock church was begun in the period 1016–20, we are looking today at a major aggrandisement of the structure, half-a-century later. The dendrochronological date obtained for the north door accords well with that hypothesis, confirming that it was contemporaneous with the limestone doorway. Bearing in mind that the 15th-century tower door is hung with three 11th-century hinges – identical to those on the north door – there must have been an equally grand west door to the church. Moreover, there was a small doorway in the north wall of the chancel (demolished in the late 18th century) that also appears to have been Romanesque (Fig. 71).

Hence, Hadstock was endowed with two large, semicircular-headed doors in the later 11th century, probably both externally covered with hide and heavily laden with decorative ironwork, which included flat-bands forming two concentric hoops. Internally, the door(s) was framed with a large bentwood hoop and four ledges, all of rounded cross-section and fixed with nails and a large number of clasping roves. The lost door from Oxford, was closely similar (Fig. 121) and the image of a large door at Hastings, depicted on the Bayeux Tapestry, confirms the design and nature of Anglo-Saxon door joinery (Fig. 125). Doors with rounded ledges, mostly fixed with clasping roves, survive at about eighteen locations in England, the majority in the east and south of the country. In date, they have nearly all been attributed to the 12th century, but some could be late 11th. These reflect the survival of long-established Anglo-Saxon joinery practices for several generations after the Norman Conquest.

While Westminster possesses the oldest iron strap on an English church door, Hadstock has the earliest surviving hinges, albeit partly renewed in 1846. The six hinges on the north and west doors comprise an exceptional group, the straps all decorated with a multitude of holes, arranged in three rows; the fixing nails are confined to the central row. A subtle difference is present: on the north door the outer rows are staggered, but on the west door the holes are in register (Fig. 139). The overwhelming majority of early hinges were provided with only a single row of nail-holes, but the south door at Hartley (Kent) has C-hinges and other straps, all pierced in two staggered rows (Fig. 140).[62] Geddes dated these to *c.* 1150–75. However, the strapwork on the door at Little Hormead, dated by Geddes to *c.* 1125–50, is even closer to Hadstock, with three rows of holes in the hinges and an intermediate strap.

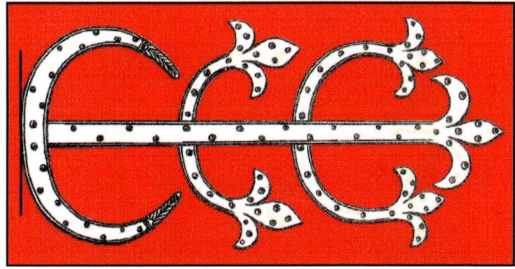

*Figure 140: Hartley church, south door. Hinge-strap mounted with three C-scrolls, two of which have fleur-de-lys terminals. In total, there are seven straps on the door, all pierced with two staggered lines of small holes, as in the north door at Hadstock. After Brandon and Brandon 1874, Sec. 2, pl. 2*

The upper and lower rows are synchronized as pairs, and the holes for the middle row are set equidistant between them (Fig. 141). Hadstock and Little Hormead are only fifteen miles apart, and there can be no serious doubt that their ironwork was produced by the same smithy.

Edge-bands made of narrow, flat-iron strips are present on some English doors, although often not complete around all four margins. At Rochester, only the top band survives, two more are attested by scars and nails, and the absence of the fourth is explained by the left-hand edge of the door having been trimmed (p. 117; Figs 93 and 96). Edge-bands partially survive on three sides of the door at Eastwood and Little Hormead (Figs 124B and 141; *cf.* also 123B). The hooped iron bands at Oxford and Hadstock are not truly edge-bands because they are set well in from the margins of the door, and the scrolled ends of the hinges of the latter lay outside the hoop (Figs 121 and 123A). The reason for this was doubtless to allow the hoop to be fully visible when the door was closed.

The remarkably plain, early Norman door at Manningford Bruce (Wilts.) has a complete iron hoop like the outer one at Hadstock, but it is also a replacement. Externally, the face of the door is divided into seven panels by horizontal iron straps (two replacement hinges, three original straps with a split-curl at one end only, and one plain strap in the arched head). Some evidence is present for nailed attachments, but much of the timber has been renewed. On the interior, the three wide boards are held together by roughly rounded ledges attached with clenched nails. The panels cry out for ornamentation, which could have been supplied by paint, either applied directly to the boards or to a layer of hide. A late 11th-century date has been claimed for the church, and Geddes suggested *c.* 1100 for the door.[63] Since it is most likely original to the present building, it would be worth dating by dendrochronology.

The late 12th-century door of two leaves at Sempringham (Lincs.) has a similarly placed semicircular edge-band.[64] This is also a characteristic of early Swedish

*Figure 141: Little Hormead church, north door. Horizontal iron strapwork pierced with multiple holes in three synchronized rows, as in the west door at Hadstock. Gibson and Hewett 1983–86, fig. 2*

doors. Another singular feature of the Hadstock door 1 is the former presence of a second complete iron hoop sited a short distance within the outer one, a detail that does not appear to recur elsewhere in England, but there is circumstantial evidence at Little Hormead that, before its truncation, the door exhibited concentric iron hoops, with a continuous decorative band between them (Fig. 123B). Again, this is a feature found in Sweden, and two examples are illustrated by Geddes, at Rogslösa and Stroja.[65] The former has a continuous band of interlace between the hoops, interrupted only by the hinge-straps.

At Hadstock, doors 1 and 2 must have been twins, or nearly so. The six hinges confirm that the doors were of similar size, and the arcature of the fragment of flat iron bar casually nailed to door 2 shows that it was part of an inner hoop like that on door 1 (Fig. 92). One of the most remarkable features of that door is the huge number of small nail-holes on the outside, indicating that it was packed with decorative motifs within the area defined by the iron hoops (Fig. 86). The eye can easily trace both straight and curved lines, but the density of nail-holes is too great to permit a convincing reconstruction of the lost decoration, which may have included geometrical patterns (*cf.* Little Hormead).[66] The band between the inner and outer hoops was most likely filled with a repeating motif, again as seen at Little Hormead (Fig. 141).[67]

On the right-hand edge of the Hadstock door, beyond the present ends of the hinge-straps, are three configurations of seven nail-holes. These groups stand out because they are the only nail-holes lying beyond the perimeter of the larger hoop.

*Figure 142: Copford church. Early 12th-century painted interior, nave and chancel, looking east. Red predominates in the polychromy of the wall-paintings. Author*

The holes define lost scrollwork closely similar to that surviving at the end of one of the hinges on door 2 (Fig. 139B). The evidence permits a reconstruction of the basic framework of the ferramenta on door 1, but not of the decorative infill (Figs 123A and 132).

In sum, the aggrandisement of Hadstock church in the later 11th century included the provision of two large, heavily decorated doors with iron banding, one of which – and almost certainly both – were covered with hide and painted red. We may suspect that these doors and their decorated masonry provided spectacular portals for the refurbishment of a church that may have been founded by Cnut. There had to be a significant historical reason for the degree of elaboration witnessed here: Hadstock was not rich or populous in the Middle Ages and it was one of the smaller parishes in north-west Essex. However, the parish and its much larger neighbour, Ashdon, together occupy a distinct geographical unit, and the straight boundary between them gives the impression of a single land-unit that was anciently subdivided.

### Elmstead, Copford and Castle Hedingham churches

These three churches in north Essex had differing origins, but all had hide coverings on their doors. Elmstead was a typical two-celled proprietary church situated alongside the hall and was seemingly not the focus of a larger settlement (Fig. 106). Little of the Norman church has survived later rebuilding and its small north doorway is of plain construction, incorporating recycled Roman materials, as frequently occurs in Essex. The door that hung here was hide-covered, clad with decorative ironwork and may initially have been the more important entrance because the hall lies a short distance to the north-west (Fig. 108). The south door gained precedence later, when a hamlet developed on the Colchester to Clacton road, one mile south of the church.[68] Nothing is known of the Norman south door.

Copford church also stands on its own, alongside the hall, and the hamlet of Copford Green lies nearly a mile to the west (Fig. 101). The medieval manor was held by the bishops of London and the church was most likely built as an episcopal chapel, *c.* 1125–30. It was a fully vaulted stone structure, lavishly decorated with wall and ceiling paintings throughout (Fig. 142). Both nave doors were hide-covered and externally elaborated with ironwork; the hide was painted bright red (Figs 104 and 105).

Castle Hedingham church lay in the market place of a small town that grew up outside the gates of Hedingham castle, built *c.* 1130. The church is a large, aisled structure, with a long rectangular chancel, dated to *c.* 1175–85 (Fig. 115).[69] The chancel was built first and has pilaster-buttresses, which are not present on the aisle walls.[70] The church has three Norman doors with decorative ironwork, potentially all products of the same smithy (Fig. 118). The south nave door is the largest and is recorded as having hide attached; it is highly likely that its northern counterpart was similarly covered. Whether the smaller chancel door was decorated remains an open question; it is a decade or two earlier than the nave, and the door construction differs slightly.[71] The earls de Vere had a hand in building the church, and are likely to have specified the decorative elaborations.

In terms of their construction, the doors of the three churches described above differ from those at Hadstock and Rochester in two respects. Elmstead is probably the oldest of the three. This must be one of the earliest appearances of two novel aspects of door construction: the first is counter-rebating of the joints between the boards, a feature that is not present at Westminster, Hadstock or Rochester, and the second is securing the boards with two tapered ledges, D-shaped in section and driven into corresponding matrices trenched into the backs of the boards (Fig. 110). As previously noted, both these forms of construction appear to be confined to the 12th century, but more scientifically dated examples are needed to refine the argument.

Little can be said about Copford. Although the boards from the south door have survived, they have been repositioned in the north doorway, and details of the primary construction cannot be ascertained. At Castle Hedingham the boards of the two nave doors have multiple counter-rebates in their jointing, but not in the chancel door (Fig. 143).

Finally, a few words need to be said about the adhesion of hides to doors. Theophilus advocated gluing them to the boards, and we have potential evidence for the presence of glue on the faces of the Westminster door. When a hide is fresh and supple, there would have been no difficulty in gluing it to the boards as a perfectly flat sheet, but over the course of decades (and centuries) of exposure to atmospheric changes, it would dry out, buckle and curl at the

*Figure 143: Castle Hedingham church, south nave door. The varied widths of the boards and the joggles are indicative of counter-rebating. Author*

edges. These tendencies would have been contained – but not obviated – on doors fitted with decorative ironwork, especially edge-bands. This is well demonstrated by the door from Högby, where the glue has long-since failed and the leather has buckled, but has been held *in situ* by the copious ferramenta (Figs 136 and 137).

However, this would not have been effective on the Hadstock and Oxford doors, where the iron banding was set well back from the perimeter: the exposed edges of unsecured hide would soon have become detached and flapping. This could be prevented by folding the hide over the edge of the door, and securing it there with something more resilient than glue. Fortuitously, the edges of the Hadstock door have not been trimmed and they display a series of nail-holes that once attached a band of either timber or metal stretched around the perimeter of the boards (Fig. 82).

## How did the 'Dane-skin' legend originate?

Twelve church doors known to have had hide coverings have been listed at eleven locations in England and Wales. Folklore has linked all but one of them to acts of sacrilege supposedly carried out by Danish invaders, although alternative origins have been advanced for the skins at Worcester and Westminster. However, the villages of Hadstock and Copford have received the most extensive mentions in county histories since the late 17th century. The villages are situated in northern Essex, 26 miles apart, and mid-way between them is Castle Hedingham, where the church also had at least one hide-covered door, and probably two.

The distribution of places nurturing 'Dane-skin' legends correlates to a considerable extent with recorded Viking activity. The Thames was a highway into southern England, and Viking camps were established in 893 on the north bank at

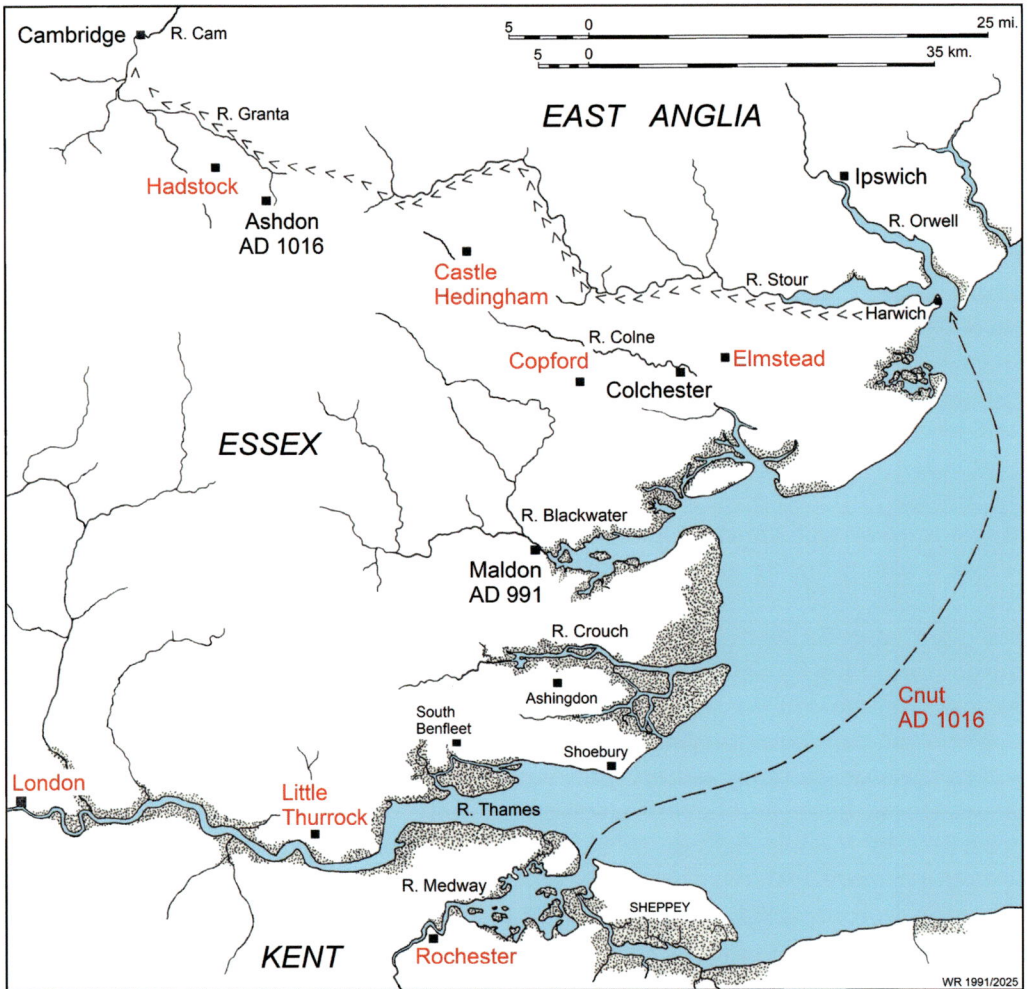

*Figure 144: Map of Essex, showing the probable route of the Danish invasion in AD 1016 and the location of the battle of Assandun (Ashdon); coastal marshland stippled. Churches with 'Dane-skins' on their doors are labelled in red. Author*

Shoebury, South Benfleet and Fulham (London). Little Thurrock church, formerly with its hide-covered door, also lies on the same bank (Fig. 144). Slightly earlier, the Viking penetration southward, into the river Medway, took them to Rochester (885), where a fortification was constructed. It was here that Pepys saw skin-covered doors at the cathedral and learned about the supposed Danish connection.

The eastern coastline of Essex is fragmented by the rivers Crouch, Blackwater, Colne and Stour, and several place-names have been popularly interpreted as having Danish associations: thus Canewdon is repeatedly referred to in local histories as recalling

Canute. Similarly, Danbury supposedly takes its name from a Danish fortification, but these are both false etymologies.[72] Allusions to Danes are present in field-names, surnames and the common names for certain plants. Although these associations have persisted for centuries, their antiquity cannot be vouchsafed.

Nevertheless, two encounters between the English and Danish armies certainly took place on Essex soil. The first was the Battle of Maldon in 991, when the Danes attacked Essex from the sea, bringing their ships into the Blackwater estuary, and the second was the inland Battle of *Assandun* in 1016. The latter marked a turning point in English history, when Edmund Ironside was decisively defeated by the Danish army, commanded by King Cnut. The Anglo-Saxon royal succession was thereby supplanted by a Scandinavian one.[73]

The site of *Assandun* has long been debated, there being two claimants: Ashdon in the north-west corner of Essex and Ashingdon in the south-east corner.[74] When the topographical and archaeological evidence is weighed objectively, it is clear that Ashdon is the only viable contender. In 1016, the Danes, having been routed from Kent, set sail from Sheppey and headed for the North Sea, with the aim of penetrating Midland England, once again beaching their ships in one of the river estuaries on the Essex coast. The most expeditious overland route would have been along the valley of the river Stour, which divides Essex from Suffolk. On a previous raid into East Anglia the army entered Suffolk via the river Orwell, which debouches into the North Sea at the same location as the Stour (Fig. 144).

A large natural harbour was formed by the Stour-Orwell estuary, guarded by the fortified promontory that is the site of Harwich today. The Danish army most likely took the route westwards along the south side of the Stour valley, until it reached the high ground in the vicinity of Ashdon, and then descended into the river valleys of the Granta and Cam, to Cambridge, and thence into Mercia.[75]

Accounts suggest that Cnut was returning via the same route from Mercia, laden with plunder, and when he reached the border between East Anglia and Essex, in the vicinity of Ashdon, he found Edmund's army, which had marched up from London, waiting to intercept him. It surely cannot be coincidence that all four churches in northern Essex claiming to have 'Dane-skins' on their doors lie but short distances to the south of the route just described between the Stour-Orwell estuary and Cambridge, namely from east to west: Elmstead, Copford, Castle Hedingham and Hadstock (Fig. 144).

Between the mid-11th and 14th centuries, when new doors were being made for churches, a small proportion of them were covered with distinctively painted animal hides. Initially, the local population would have been well aware of the true nature of these coverings, but as the centuries passed and the skins became dilapidated and curled, their original decorative function would have faded from local memory, leaving ample room for grisly speculation. Essex had suffered multiple Viking incursions between the 9th and early 11th centuries, and these were profoundly embedded in folklore. The temptation to link the remains of these ancient hides, for which there seemed to be no other ready explanation, with the Danes is understandable. With recent advances in genetic studies, we are finally able to identify the skins as cow and horse or donkey, just as Theophilus stated nine hundred years ago.

As to when the 'Dane-skin' legend was initiated, there is no direct evidence, but folklore had fully embraced it by the 17th century. The notion that the hides were human skin clearly appealed to a credulous sector of the population because it was enthusiastically repeated through generations of guidebooks and county histories. Whether the gory legend was born and incubated in northern Essex, and spread from there to other localities with hide-covered doors and a history of Viking incursion cannot be definitively established, but there is circumstantial evidence. Either way, Hadstock appears to have achieved the greatest notoriety in respect of its 'Dane-skin' door.

Exactly where the critical battle took place in 1016 has yet to be established, but it was on the Ashdon-Hadstock hilltop. Archaeological activity may one day relocate the battle site: somewhere in the Ashdon area, the dead from both the English and Danish armies must surely have been buried? *Ignoramus sed non ignorabimus.*[76]

# Notes to chapters

## Notes to chapter 1 (pp. 1–19)

1 Evans 2000, 19.
2 *E.g.* Howard & Crossley 1917, chap. 2.
3 Cecil Alec Hewett (1926–98), craftsman and historian of carpentry.
4 Hewett 1974; 1980; 1985.
5 Addyman & Goodall 1979.
6 Geddes 1999.
7 Miles & Bridge 2012; new lists are published annually in *VA*.
8 Geddes 1999, 13–14.
9 Pepys 1970, 2, 70.
10 A pair of doors with three hinges apiece and some nondescript scrollwork was drawn by Daniel King (1672, pl. 9). The depiction is too crude for meaningful discussion.
11 Baxter 2006.
12 Geddes 1999, 362, with references.
13 Barnard 1931, 104, referring to Prattinton ms 02/08, p. 627.
14 The backs of some doors have subsequently been flattened by replacing the ledges with a layer of boarding: *e.g.* Castle Hedingham church and Rochester Cathedral ('Gundulf's door', chap. 6).
15 Way 1848, 186; Brooks & Pevsner 2007, 677.
16 Geddes 1999, 29, fig. 2.17.
17 Brassington 1894, 83.
18 I am deeply indebted to Fiona Keith-Lucas (Cathedral Archaeologist) for tracking down this important item, and providing information about it.
19 Accn no. 288. Way 1847, 46. Also cited in *Notes & Queries* 4th ser. 5 (1870), 310–11. Some accounts refer to the skin as being that of a Dane.
20 Hunterian Museum, accn RCSMS/Quekett/Division 2/Hc 113.
21 *Proceedings Cambridge Antiquarian Society* 11 (1904), 179.
22 Brassington 1894, 83.
23 The present small doorway in the south wall was a temporary access point in the 1250s (Fig. 19, door 3), blocked with masonry and only reopened in 1871 (Rodwell 2010a, 3).
24 Dart 1723, I, 64. Dart's plan (p. 69) shows the doorway in the south wall of the transept, but it does not illustrate the revestry beyond. The latter is now part of St Faith's Chapel.
25 Ackermann 1812, 2, 26.
26 'Proceedings' of the [Royal] Archaeological Institute, 1 Apr. 1853, *Archaeological Journal* 10 (1853), 167–8. This was an erroneous *addendum* to a previous paper (Way 1848).
27 This archaic term, derived from the Latin verb *cancellare*, was introduced in the 1650s to describe something that had a latticed or net-like structure.
28 Stanley (1882, 369) erroneously stated that all three doors to the 'revestry' were skin covered. His error was repeated by Tyack (1898, 164–5).
29 Swanton (1976, 24) and others have confused the long-lost sacristy door with the one that still hangs in the chapter house vestibule.

30  Large church doors were commonly up to 7.5 cm (3 ins) thick, and attracted no comment.

31  In the early 18th century, the southernmost bay of the transept was enclosed by a timber screen, and its door was probably that to which Dart alluded.

32  The door is first glimpsed in an engraving by J.S. Storer, 1805; it is also seen, standing partly open, in a view by J.P. Neale, 1815 (Nightingale 1815, opp. 93). Perkins (1952, 29) incorrectly attributed the door to Gilbert Scott.

33  The lights were 'open', but infilled with boarding in the mid-20th century.

34  This may have been the work of Henry Keene (Surveyor of the Fabric, 1752–76).

35  Scott 1860; 1861, 37–8.

36  Stanley 1868, 384.

37  Letter dated 15 Mar. 1853. Transcribed in Reynolds 2011, 39, n1. Later references to this location use the terms 'skin cellar' and 'skin door'.

38  Brindle 2010.

39  Scott 1860; 1861, 40.

40  Harrod 1870, 377 (copied from Stanley 1868, 384).

41  Brassington 1894, 83n.

42  Noppen 1936, 33.

43  Stukeley 1724, 75; 2nd edn 1776, 79.

44  BL, Add. 5836, fol. 17, dated 20 Oct. 1726.

45  Morant 1768, II, 543.

46  Cromwell 1819, II, 147. The drawing is dated 1818.

47  BL, Add. 6744, fol. 3. The drawing is signed 'JAC Del. 1809'.

48  Wright 1836, II, 103.

49  Coller 1861, 543.

50  Tyack 1898, 163.

51  Gray 1906–07, 223.

52  Richard Neville, FSA, 1820–61.

53  *Proceedings Cambridge Antiquarian Society* 11 (1904), 179. This piece of skin was given to Mr Deck's father by the rector, the Rev'd C. Townley.

54  *I.e.* £3.15 in modern currency. Stevens's Auction Rooms, 10 Jan. 1905. Gray 1906–07, 224.

55  The museum was opened in 1835, but its accession register was compiled in 1847 and does not record the source of the Hadstock items.

56  Morant 1768, 192.

57  Newcourt 1710, II, 191.

58  Muilman 1772, VI, 176.

59  Wright 1836, I, 406.

60  Cromwell 1818, I, 64.

61  Coller 1861, 554–5.

62  Benton 1936.

63  Hewett 1974, 99, fig. 68; Geddes 1999, 311, figs 3.8, 4.115, 4.116. The south nave door was the only one protected by a medieval porch, and it is not surprising that the hide covering survived here long after it had been lost from the north nave and south chancel doors.

64  Tyack 1898, 160. His reference to doors (plural) is correct because there are three 12th-century doors in the church (Hewett 1974, 99, fig. 68).

65  RCHME 1923, 86–7.

66  Quekett 1888, 117. A transcript of the relevant part of Quekett's autobiography was published by Robert Pierpoint of Warrington: *Notes & Queries* 10th ser., 1 (1904), 73–4. The incident occurred in 1848, and hence the discovery was not associated with the 1878–84 restoration.

67 Gray 1906–07, 225.
68 'Blessed are the possessors', a Catholic term.
69 Enquiries directed to Thurrock Museum and Local History Society did not elicit any further information.
70 Addyman & Goodall 1979; Geddes 1999, 14, 373, figs 2.13, 4.23, 4.24.
71 *The Treasury Magazine* 10 (1908), 417; Thomson 1955, 128ff; Addyman & Goodall 1979, 103–4. The skin was trapped under the ironwork of the ship.
72 RCHME 1934, 162, pl. 45. Door not included in Geddes 1999.
73 Hewitt 1901–02, 142, pl. 28.
74 Museum accn no. 1686. Regrettably, it was not possible to record the genetic composition of this piece in time for its inclusion in this volume.
75 Gray 1906–07, 228, 229.
76 *Somerset County Herald*, 15 May 1943, p. 4, col. 4. The tower and west doorway date from 1487, but the south nave doorway is 12th century.
77 *The Treasury Magazine* 10 (1908), 417.
78 Hughes 1949, 69.
79 Noted in 1971 by A.J. Taylor, who was informed by Peter Curnow, a historic buildings inspector with the Ministry of Works. See also Bettley *et al.* 2019, 284.

## Notes to chapter 2 (pp. 21–36)

1 Way 1848.
2 *Ibid.*, 185.
3 Pepys 1970, II, 70.
4 The appellation 'Pyx door' occurs in Hewett 1978; Rodwell *et al.* 2006; Bridge & Miles 2012, 76–7.
5 Otherwise known as Richard (or Robert) de Podelicote.
6 Stanley 1868, 383–5.
7 Harrod 1870, 377. Although there has been a long-held belief that the 1303 robbery took place in the Pyx Chamber, that is not certain.
8 Stanley, in later editions of his *Historical Memorials*, added various embellishments, including stating that the three doors in the sacristy were skin-covered, whereas only one was (Stanley 1882, 369).
9 Way 1848, 186; Quekett 1849, 152–3.
10 Quekett 1849, 155, pl. 24.
11 A single slide of hair from the Worcester Cathedral hide survives, and is held by the Royal College of Surgeons.
12 Quekett 1849, 155–6.
13 Way 1848, 186.
14 Scott 1861, 40.
15 *E.g.* by Harold St George Gray (1906–07).
16 Catalogue of the Hunterian Collection of the Royal College of Surgeons, London, 1923.
17 RCHME 1924, 79b. Lethaby (1925, 129) mentions the door *en passant*: 'is that which has fragments of some skin (said to be human), which lined it within'.
18 Quekett 1848; 1849.
19 Only his slide collection appears to have survived.
20 Tyack 1898, 163.
21 Sir Benjamin Brodie, surgeon and physiologist, 1783–1862.
22 Quekett 1849.

23  Way 1848.
24  Tyack 1898.
25  Gray 1906–07.
26  Swanton 1976.
27  *Ibid.*, 22.
28  Swanton 1976, 25. The owner of the private collection was D.M.D. Thacker. It is not recorded how he came into possession of this piece of skin, or what subsequently happened to it.
29  WAM Z/4, Papers of Arnold J. Taylor. He was Chief Inspector of Ancient Monuments at the then Ministry of Works (later, Department of the Environment), 1961–72. Since the chapter house and Pyx Chamber were in his department's care, he had a *locus* for studying the vestibule door. Subsequently, 1979–92, Taylor served on the Abbey's Architectural Advisory Panel.
30  WAM: letter 29 Jan. 1967.
31  Letter to Taylor, 10 Aug. 1967; Stockholm Museum Inventory no. 4779.
32  Westminster Abbey Collection. It measures 12 × 6 mm. The outer surface is smooth and cream in colour; no trace of paint is visible. The inner face preserves the impressions of wood-grain, and there is a hint of rust-staining from an iron fitting.
33  WAM: letter 15 Feb. 1967.
34  WAM: letter 27 Nov. 1967.
35  WAM: letter 10 Jan. 1968.
36  WAM: letter 15 Mar. 1968.
37  Swanton (1976, 25) erroneously stated that the samples were from 'the old revestry door', not the vestibule door. These samples are held in the Westminster Abbey Collection.
38  Reed 1972, 184–7.
39  *Ibid.*, 187.
40  *Ibid.*, 285–7.
41  Swanton 1976, 26.
42  Accn no. COLEM.1926.5171 (examined, 24 Oct. 2024). It was gifted by A.F. Nicholson in 1925, whose connection with Copford has not been elucidated. A second piece of skin (not seen by the present writer) is unaccessioned and was donated in the 19th century by Canon Peter Wood, Rector of Copford, 1861–78. It was given to Wood by Theobald, the parish clerk, who said that the skin was taken from beneath the ironwork of the door. Noted by Gray (1906–07, 222).
43  Swanton 1976, 26. Also correspondence with David Clarke, Curator, Colchester Museum (dated May 1973; Colchester Museum Archives).
44  *Notes & Queries*, 4th ser., 5 (1870), 310–11; 10th ser., 1 (1904), 155. It was illustrated by Gray (1906–07, pl. following 222). It does not have a similar appearance to the other Copford pieces, and no explanation is recorded as to why Quekett donated the skin to Taunton Museum. He was born in Somerset, and bequeathed a collection of antiquities to the Somerset Archaeological Society; consequently, it is possible that the piece of skin was a misidentified local discovery (Gray 1906–07, 222).
45  Report dated 14 May 1974. Cited in part by Swanton (1976, 26).
46  *Ibid.*, 26.
47  Geddes 1999, 15.
48  Research Fellow, All Souls College, Oxford.
49  Macleod, R., Evans, M., Geddes, J., Rodwell, W., Wingfield, C., Keith-Lucas, F., Munoz-Alegre, M.M., Gilbert, M.T.P., and Collins, M.J., *in prep.*
50  Swanton 1976.
51  *Ibid.*

52 Report dated 14 May 1974.
53 Fiddyment *et al.* 2015.
54 Brandt *et al.* 2023.
55 Miranda Evans, Jane Geddes, Warwick Rodwell, Carolyn Wingfield, Fiona Keith-Lucas, Marta Munoz-Alegre, Tom Gilbert (who first piqued my interest in Dane-skins) and Matthew J. Collins, to all of whom I am indebted for their contribution to this research.

## Notes to chapter 3 (pp. 37–66)

 1 Rigold 1976, 13.
 2 Hewett 1978, 214, figs 14, 15. See also Hewett 1980, 26, figs 23, 24; 1985, 155, fig. 149.
 3 Geddes 1999, 22–3, 344, fig. 2.5.
 4 *E.g.* by Campbell & Tutton (2020, 38) in their volume on the history and conservation of doors.
 5 For a fuller description and illustrations of the cloister and vestibule, see Rodwell 2010a, 1–8.
 6 The day-stair was demolished *c.* 1591 and replaced with a timber staircase to the library: Rodwell 2010a, 4.
 7 Hewett (1985, 155–6) incorrectly stated that the planks were quarter-sawn.
 8 The peg in board 4 is ancient. The position of the peg in board 5 is now occupied by a nail.
 9 Compare Hewett 1978, fig. 14, and 1985, fig. 149.
10 *E.g.* Miles & Bridge 2010a, 254.
11 Underwood 2001, 78–80.
12 The layer of hide is 1 mm in thickness and can be seen along much of the top edge of the hinge.
13 At that time, the Pyx Chamber and vestibule office were under the control of English Heritage. I recall being shown the roll in my youth.
14 The strip has never been part of the Abbey Collection. Enquiries to English Heritage failed to elicit any information.
15 This analysis was conducted on the strip of hide that still remains trapped between the mid-height iron band and the boards. Hence, there can be no doubt that this represents the original hide covering of the external face.
16 The nail-holes are not visible on the south face of the door, having been covered by Scott's new strap-hinge.
17 WAM Z/4. The pieces of hide are pierced with holes the size of dressmakers' pins, indicating that they had been displayed, probably in the Undercroft Museum in the east cloister, sometime before its refurbishment in 1986.
18 Westminster Abbey Paintings Conservator.
19 Report dated 5 Dec. 2024.
20 Cotte *et al.* 2016.
21 Bruce-Mitford 1975, 354–65. There, the bolts are referred to as 'plank-joint rivets'.
22 There are three double-drilled holes in the upper curve of the scroll, and three in the lower curve. A single hole on the very edge of the door, associated with the lower group, is all that remains of a fourth double-drilling.
23 *E.g.* in board 3, one of the clench-bolt holes for the lower hinge-strap has a second drilled hole 20 mm above it; and in the same board, below the mid-height iron strap, are two drilled holes 35 mm apart. No explanation can be offered for these anomalies.
24 Sheets of canvas were purchased in that year, to hang in the window openings, pending the installation of the glazing (Carpenter, in Rodwell & Mortimer 2010, 34).
25 Rodwell 2010a, 23–4.

26 It tapers from 65 mm to 52 mm.

27 The lower pintle is possibly a Victorian replacement but it is plastered all around, concealing the leading.

28 This is a common type of bolt on doors and chests, and the hasp may either be padlocked to a staple (as here), or engage with a fixed lock. The latter is exemplified on the south porch door at Saffron Walden (Geddes 1999, fig. 6.60).

29 In *c.* 2013, when a modern drain became blocked, the south cloister flooded from end to end.

30 Dean & Hill 2014.

## Notes to chapter 4 (pp. 67–81)

1 Paul Reed and Peter Massey, master carpenters.

2 This chapter is an abridged and slightly revised version of an account first published in *Regional Furniture* (Massey & Reed 2021).

3 Rodwell *et al.* 2006, 25–7; Rodwell 2009; 2012a, 44–6, with a detailed description of the door.

4 Harvey 1987, 294–5. The term is no longer considered to refer expressly to a carpenter, but to the overseer of the building works. For the current interpretation of 'church-wright', see Gem 2009, 169–70.

5 Miles & Bridge 2005, 8.

6 Rodwell 2012b, 145.

7 Cheddar: Rahtz 1979; Yeavering: Hope-Taylor 1977; Cowdery's Down: Millet & James 1983; Bishopstone: Thomas 2010; Lyminge: Thomas 2013; Dublin: Murray 1983; York: Hall 2014.

8 Goodburn 1992.

9 Goodburn 2007, 46–89; reconstruction in the Museum of London.

10 The earliest account of Anglo-Saxon carpenters' tools was published by Sir David Wilson (1976, 253–9).

11 Geddes 1999, 19–29, 50–7; Hewett 1982, 78–94; 1985, 155–87.

12 This was not always the case because some hinges were nailed straight through the boards and into the tapered ledges: *e.g.* at Little Hormead. Perhaps seasoned boards were used for making these doors, so that significant shrinkage would not occur. Once the tapered ledges had been driven home and external ironwork nailed on (pinning the ledges), there was no option for tightening-up. [WR]

13 Hewett 1982, 78; Geddes 1999, 20, table 2.1 lists doors with rounded ledges.

14 Hewett 1985, 162, fig. 156; Geddes 1999, 29, fig. 2.16. Both illustrate counter-rebating.

15 Hewett 1985, 161, fig. 155; Geddes 1999, 24, table 2.4 lists doors with lattice bracing.

16 Goodburn 2007, 302–15; Hill & Woodger 1999, 47–51.

17 Milne 1992, 106–14; Hall 2014, 775, quoting Goodburn.

18 Darrah 1982, 219–23.

19 Bridge & Miles 2012, 77.

20 Geddes 1999, 21, table 2.1; Harrison 2007, 53, and pers. comm.

21 Goodburn 2007, 302, and in correspondence with the authors.

22 Dunning & Goodman 1959; Goodman 1978, 55. The plane is in Maidstone Museum.

23 Williams 2011, 41–3, pl. 7, fig. 8. The plane is in Maidstone Museum.

24 Being at the door's extremities, boards 1 and 5 only required dowels in one edge. The four dowels in the lower row are all at the same distance above floor level; those in the upper row are not aligned.

25 Hull Museum: accn KINCM 1984.58.3. Excavation by J. Dent, 1982; note in *Antiquaries Journal* 63 (1983), 387.

26 Ottaway & Cowgill 2009, 263, fig. 7.8.

27  Hewett 1978, 214; 1980, 26; 1982, 343.
28  Arwidsson & Berg 1982, item 57, pl. 13.
29  Hewett 1978, 214–16; 1980, 25, showing pegs end-wedged; 1982, 344, the same drawing corrected, with the pegs unwedged.
30  Most of the pegs appear to be modern replacements, and only two or three are likely to be medieval. [WR]
31  Items were joined and jewels secured in the Staffordshire Hoard with animal glue containing beeswax (Fern *et al.* 2019, 137). Theophilus described animal and cheese glues for use in carpentry in the early 12th century (Hawthorne & Smith 1979, 26–7).
32  Underwood 2001, 78–80; Rodwell 2012a, 45.
33  Goodman 1964, 123; Arwidsson & Berg 1982, saw no. 42.
34  Hewett 1980, 21–3; Bridge & Miles 2012, 73–4. Traces of red paint were recorded on the hide.
35  Geddes 1999, 27; Harrison 2007, 53.
36  Hewett 1980, 46, 57, fig. 41; Miles *et al.* 1999, 5.
37  Hewett 1980, 211, 213; Rodwell 2012a, 22; 2012b, 147. Hewett incorrectly described the stiles as 'chamfered'. In 1979, I dismantled the post-medieval glazing in the north-east window and found that it was set into a rebate, crudely cut with a chisel while the frame was *in situ*. [WR]
38  Goodburn 2007, 302. Darrah (2009, 98) included three other tools: chisel, adze and twybil (a short-handled, T-shaped, double-headed tool, used for chopping out mortices).
39  Hull Museum: accn KINCM 1984.58.3 (Skerne split-socketed chisel).
40  Rogerson & Dallas 1984, 77–8, fig. 117.13. Norwich Castle Museum: accn. NWHCM 1950.12.887 (Thetford saw).
41  Arwidsson & Berg 1982, saw no. 41.
42  Hewett 1982, 339–41. In a report written in 1978, Hewett did not mention saw kerf-marks: Cronyn & Horie 1985, appendix 7, 65–7. This evidence needs reassessing.
43  Hewett 1978, 214; 1980, 26; 1982, 343 in McGrail 1982.
44  Correspondence with Christine Rauer, Reader in Medieval English, University of St Andrews, Scotland.
45  Although not evidenced by the Westminster door, other common woodworking tools have survived from the period, including gouges: there is one in the Nazeing hoard (Morris 1983, 34, fig. 3e) and a socketed example in the Crayke hoard (Sheppard 1939). [WR]
46  James 2012, 7.

## Notes to chapter 5 (pp. 83–110)

1  Rodwell 1976.
2  William Cole mss. BL, Add. 5836, fol. 17. Confirmation of the existence of this door in 1720–30 is recorded in William Holman mss, Essex Record Office: D/Y1, T/P 195/16. The chancel subsequently fell into ruin, was demolished and rebuilt in 1792.
3  Tweddle *et al.* 1995, 211–12.
4  The recessed area is larger than the door, in both width and height, but there is no eastern edge to it, which was probably lost when the transverse wall between the nave and crossing was demolished in the 13th century.
5  Neville 1847, 34–5.
6  Soc. Ants, Minute Book, vol. 23 (1789–90): SAL/02/023.
7  Museum register, pp. 198D–E. Accn SAFWM: 1847.4/1 (skin) and 1847.4/2 (iron). The name 'W.G. Gibson?' is noted, but he has no known connection with Hadstock.

8   The register erroneously records that the drawing was made by 'GNU', but it is clearly initialled 'GNM' (Geddes 1999, fig. 4.8). G.N. Maynard was the museum curator 1880–1904. The illustration could not be located in 2025.

9   For the political background, see Lawson 1993, 9–48.

10  *Ibid.*, 142.

11  Robert Miller Christy (1861–1928) was a well respected and prolific author, but sometimes strayed too close to the world of fantasy. For his biography and publications, see *Essex Journal* 42 (2007), 23–6.

12  Christy 1925, 188.

13  *E.g.* RCHME 1916, 143–5; Baldwin Brown 1925, 365–8; Clapham 1930, 99, 126; Cobbett 1937; Taylor & Taylor 1965, 272–5, 1080; Fernie 1983a, 169–70. The last classed Hadstock as belonging to the Saxo-Norman 'overlap'. Surprisingly, Fisher (1962, 302) did not include Hadstock amongst the 'greater Anglo-Saxon churches'.

14  Prof. Eric Fernie and Dr Richard Gem. See Fernie 1983a, 72; 1983b.

15  BL, Add. 5836. His description of Hadstock begins on p. 18v.

16  *Ibid.*, p. 18v.

17  For Hadstock, see BL, Add. 6739, 6744, 6756, 6768.

18  BL, Add. 6768, fols 89–90. The sketchbook is a compilation of loose sheets and is signed on the flyleaf 'T. Kerrich M.C.C. 1797'. The description of Hadstock church is accompanied by a small sketch of the north doorway (omitting the door). In the index (fol. 301), Hadstock falls under the heading 'Mr Essex's MSS. Miscellanies, Vol. 1.' The notes in this volume are by several different hands.

19  BL, Add. 6744, fol. 3. The last digit of the date is not entirely certain. This drawing is pasted in a volume entitled 'J. Essex, Architectural Collections'. It comprises 52 folios of church details, drawn in ink by many different hands: a few bear names or initials, which do not include those of James Essex. The style of the Hadstock drawing, and the fact that it is in pencil, cause it to stand apart from all the other material in the volume.

20  BL, Add. 6756, p. 280. This is pasted into Kerrich's scrapbook. The initials, which are clearly written, are not those of James Essex. The inferior quality of the sketch is also incompatible with Essex's work.

21  Clarke lived at Roos Farm, Debden Road. He is not to be confused with Joseph Clarke, FSA, architect (1819–88).

22  The superscript location and underlining of the lower-case 'h' indicates that it was not a middle initial, but associated with the letter 'J'. Joseph seemingly adopted the abbreviation 'J$^h$' to distinguish his initials from those of his brothers Joshua and John.

23  Accn LDSAL 2020.1.183. The illustrations are in sepia ink, drawn on one sheet of paper (fol. 88; 255 × 192 mm), which bears the watermark 'G YEELES 1826', showing that it was a product of Bathford Paper Mill (Bath, Som.); the text is written on both sides of a second sheet (fol. 78), and the legibility of this document is poor. Also, the sheets have been folded multiple times, to fit them into a rectangular space, having the proportions of a small envelope.

24  Player 1845, 64.

25  BL, Add. 36433, no. 601; Buckler, Architectural Drawings, vol. LXXIX, misc. vol. IV.

26  Hewett 1974, 97, fig. 65.

27  Hewett 1978, 211–13, figs 11, 12. Republished in Hewett 1980, 21–2, figs 19, 20.

28  Geddes 1999, 21, 53, 327–8, figs 2.1, 2.11, 4.6–4.8.

29  The event was organized by The Hadstock Society, and the participants who presented papers were David Andrews, Martin Bridge, Patricia Croxton-Smith, Eric Fernie, Jane Geddes, Adrian Gibson† and Warwick Rodwell.

30 Geddes 1999, fig. 4.8.
31 The second paragraph is evidently a quotation from an untraced source.
32 The widths of boards 1–4 are 36, 36, 40 and 56 cm, respectively; no edge-dowels.
33 These have been superseded by Victorian bolts.
34 Ledge C. Ledge B is 45 mm wide. The hoop is more slender, and on the arch its width tapers to 35 mm.
35 Interspersed were a few odd fixings, more like pipe-cleats than roves: they have no hole for a nail.
36 Hewett 1980, 21; Geddes 1999, 28. Hewett asserted that the roves were 'so elongated as to encircle the wood and prevent it splitting when the clenches were formed': the nails were not clenched.
37 It is not known when the shell-gimlet was invented, but it is recorded from the 15th century onwards. It is a small drill with a T-bar, twisted by hand, and is a tiny version of the shell-auger, a fundamental pre-Norman tool.
38 The new boards were not rebated, have butt-joints, and shrinkage has opened gaps between them.
39 The middle and lower straps are depicted slightly higher than they should be, the bases of the nook-shafts are incorrectly shown and the recess for the stoup is wrongly positioned.
40 The return on the rear face is 32 cm.
41 There is no historic precedent for these terminals.
42 Both scrolls are flanked by a pair of additional nail-holes, which must relate to small decorative adjuncts attached to them (Fig. 139A).
43 The rebates only needed to be 15 cm long to accommodate the back-straps of the hinges, but the carpenter made a more generous allowance; the hinges were perhaps not to hand at the time.
44 No corresponding nuts are visible on the interior and are presumably concealed under the ledges. Replacement clasping roves were fitted to the latter, and must be attached with short nails because the iron straps are directly under the ledges.
45 Geddes 1999, 364.
46 In the band between the inner and outer arched heads several groups are discernible, each of four nail-holes in roughly trapezoidal form.
47 The present ring is modern, but the diamond-shaped backplate is older.
48 The vertical elements of the hoops are composed of three separate sections, each being fitted between the hinge-straps.
49 The dimensions of the externally visible part of the door are 2.40 × 1.22 m.
50 It was briefly noted by RCHME (1916, 145): 'W doorway of plain oak boards, probably 13th century, with remains of old iron-work'.
51 Rodwell 1976, 65; Fernie 1983b, 72; Geddes 1999, 54, 327, fig. 4.9.
52 BL, Add. 6744, signed 'JAC Del. 1809'.

## Notes to chapter 6 (pp. 111–119)

1 By Jill Atherton (*olim*, John Atherton Bowen). The drawings were commissioned by the Friends of Rochester Cathedral.
2 Miles & Worthington 2002, 82.
3 Details of the alterations are described by Geddes (2006, 54–5).
4 Geddes 2006, 54. She suggested that there could have been a central rack-bolt. This was a rack-and-pinion security device that simultaneously operated two vertical bolts, one penetrating

the floor, and the other the head of the doorway, but there are too many holes, and they are not in the right positions.

5 Geddes 1999, 363; Miles & Worthington 2002, 82.
6 The board is also damaged by splits and the top right-hand corner was trimmed off when the door was adapted to fit in an arched opening.
7 Geddes 1999, 360, fig. 3.11.
8 Boards 2 and 4 are narrow and of similar width (20 cm); board 3 is wider (30 cm). Board 1 falls between these dimensions, but if it too was 30 cm, before trimming, the overall width of the door would have been 80 cm and the hinges would have fitted.

## Notes to chapter 7 (pp. 121–127)

1 Captain Anthony W.G. Lowther (1901–72), architect and archaeologist.
2 Dr John H. Harvey (1911–97), architect and architectural historian.
3 WAM Z/4. Letter from Harvey to Fletcher, 21 Dec. 1976.
4 WAM Z/4. Letter of 1 Aug. 1972. A photographic record had already been made by the Ministry of Works, 3 Feb. 1967: ref. G11.467/1–5.
5 WAM F/11500. Letter from Fletcher to R.G. Phillips (Historic Buildings & Monuments Commission), 11 Dec. 1980.
6 It was a Bank Holiday (25 August) with a trickle of tourists visiting the church, but that did not impede progress. Fletcher stood on a trestle to reach the top of the door, where he measured the ring-widths, one-by-one, using a graticule, and called out the figures to Kirsty Rodwell, who logged them. Meanwhile, Hewett and I continued our investigation and discussion of the church's carpentry. He drilled two cores for potential dendro-dating of the nave roof (unsuccessful).
7 Rodwell 2002.
8 This was public knowledge and the press periodically reported on the issue.
9 He was instructed in the 1970s by the Historic Buildings & Monuments Commission (precursor of English Heritage) to attempt the dating.
10 The data are preserved in the Fletcher archive at Oxford.
11 This was initially developed for work on the medieval doors at the Tower of London, commissioned by the Historic Royal Palaces Agency.
12 Bridge & Miles 2012.
13 Plant 2006.
14 Geddes 1999, 132–4, 362–3, fig. 4.206; 2006. A grant was obtained from the Society of Antiquaries to facilitate dating by dendrochronology.
15 Miles & Worthington 2002, 82; Geddes 2006, 54.
16 Again, this work was funded by a grant from the Society of Antiquaries.
17 Miles *et al.* 2004, 98.
18 Adrian V. Gibson (1931–2006), architectural historian and timber-framed building specialist.
19 Miles & Bridge 2010a, 254–5. The reliability of the correlation between the tree-ring curves from the measured boards and diverse oak chronologies and datasets from Northern Europe is calculated using the Student's *t*-test (*t*-value).
20 The samples were taken by Dr Daniel Miles, assisted by Dr Martin Bridge, with Dr Jane Geddes also present and recording observations.
21 For the full technical report, see Miles & Bridge 2005; Miles *et al.* 2005, 91. For additional discussion, see Miles & Bridge 2010a; Rodwell *et al.* 2006.

## Notes to chapter 8 (pp. 129–145)

1 RCHME 1922, 768; Tristram 1944, 115–19; Bettley & Pevsner 2007, 305–6.
2 RCHME 1922, 77.
3 Geddes 1999, 316, fig. 4.118.
4 The degree of weathering at the top of the door confirms that it was not truncated in the 19th century.
5 Details of these cited works by Morant and Newcourt are given in chapter 1.
6 Dr Henry Laver (1829–1917), Colchester.
7 *Transactions of Essex Archaeological Society*, new ser. 3 (1889), 94.
8 Tyack 1898, 163.
9 The overall dimensions of the door (excluding the modern framing) are 2.25 m high by 1.08 m wide. The widths of the boards are 12, 37, 23 and 34.5 cm.
10 The pegs are 10 mm in diameter and largely obscured by limewash, but eight or nine are showing through in each row. The lower row is 35 cm above the bottom of the door, and the upper row 1.15 m. This may represent the middle ledge out of three; the pegs for the top ledge may be hidden by the upper hinge-strap.
11 Mounted in a glazed and sealed picture frame; not opened.
12 The outer face cannot be seen. Fragments (ii) and (iii) are mounted in a glazed picture frame. A mid-19th-century label written in sepia ink reads: 'Pieces of Human Skin supposed to be Dane's | Found on the Church Door'.
13 Accn COLEM:1926.5171.
14 BL, Cotton Nero C.IV, fol. 17r. See Geddes 1999, figs 4.117, 4.118.
15 Geddes 1999, 199–200, 297, fig. 5.116.
16 The Norman door is now displayed on the west wall of the nave, but is demountable.
17 Benton 1936, 126–30.
18 As seen from the front of the door, boards 1–3 measure only 25.5 cm across, but their full width also includes the lap that can only be seen from the rear.
19 Only traces of the lower joggles survive, owing to decay and the door having been reduced in height.
20 There is a single, much smaller, drilled hole in the fourth board that may have held an iron rivet, but not a wooden peg.
21 Hewett 1974, 99, fig. 69. It was based on a hasty and somewhat inaccurate sketch.
22 Geddes 1999, 322, fig. 4.86.
23 Seven of the eleven nails in the hinge-strap project, and eight of the nine nails in the C-scroll.
24 RCHME 1916, 47–57.
25 Geddes 1999, 102, 311, figs 3.8, 4.115, 4.116.
26 Two are backed with Victorian diagonal boarding (as at Copford), and the chancel doorway is internally blocked with masonry.
27 Hewett 1988, 377, fig. 4. I was unable to locate evidence to verify Hewett's claim.
28 Smears of black gloss paint are associated with modern decoration of the ironwork.

## Notes to chapter 9 (pp. 147–182)

1 Tyers 1996; *VA* 28 (1997), 142, no. 29.
2 Geddes 1999, 330, fig. 4.83.
3 Dodwell 1961; Hawthorne & Smith 1979, xvi.
4 Hawthorne & Smith 1979, 26.

5 'Shave-grass' is one of several popular names for 'horsetail'. It had multiple uses, medicinal and practical.

6 Hawthorne & Smith 1979, 27.

7 *Ibid.*, 27–9.

8 British Museum: accn 1912: 0723.

9 Arwidsson & Berg 1982.

10 Christensen 1982, fig. 18.2.

11 Instead of dowels, loose tenons could be substituted, which are even stronger. They are short pieces of timber, rectangular in cross-section, inserted into matching mortices in both boards. This technique is evidenced in a door of Kempley church, dated by dendrochronology to 1114–44, and calibrated to 1128–32 (*VA* 39, 2008, 133).

12 Hewett 1988, 375, fig. 1 (incorrectly captioned as the north door); Geddes 1999, 334.

13 Addyman & Goodall 1979, 97–100, figs 19, 20.

14 Gibson & Hewett 1983–86, fig. 2. There is nothing to suggest that the bottom of the door has been truncated, to account for the loss of a second joggle.

15 Geddes 1999, table 2.6. There are doubtless more doors with this distinctive jointing, but Geddes's book deals only with those that bear decorative medieval ironwork. A hitherto unrecognized example was reported by Hugh Harrison in 2000 at Bristol Cathedral (Bettey & Harrison 2004).

16 The door remains in its original position at the head of the night-stair in St Augustine's Abbey, founded in 1140; it became Bristol Cathedral in 1542.

17 Medieval diocese of Lund, then in Denmark. The door is now in the collection of Lund University Historical Museum. Melin 2018.

18 Melin 2018, 254–5, fig. 6.

19 *Ibid.*, 257, fig. 8.

20 Ridgeway 2009, 11; Bridge & Miles 2012, 76, fig. 2.

21 Horsman 1988, 89–91, fig. 84. The upper end of the door was not accessible to record.

22 Geddes 1999, 313, fig. 2.3.

23 Hewett 1974, 100, fig. 70.

24 Geddes 1999, 372–3, fig. 2.12.

25 BL, Add. 36433, fols 665, 668. Geddes 1999, 105, figs 2.2, 4.125. John Chessell Buckler was active in the Oxford region in the 1840s and 1850s, which is when he probably saw and drew the door from St Peter's Church. Unfortunately, he seldom dated his sketches.

26 The Hadstock door, measuring 2.87 × 1.45 m, is of the same order of magnitude, but differs slightly in its proportions.

27 Hewett 1974, 98, fig. 66; Geddes 1999, 306, fig. 4.91.

28 Miles & Bridge 2010b, 103. All five boards were measured, yielding compatible results.

29 Geddes 1999, 344–5, table 2.1. I have excluded the stair-turret door in the north transept of Westminster Abbey, which Miles & Bridge (2005, 2–3) pointed out was erroneously listed: its ledges are bevelled, not half-round.

30 Geddes 1999, 19, 22, table 2.2. She also included Heybridge (Ess.), but according to Hewett (1974, 100, fig. 70), the ledges are neither rounded nor dovetail-wedged: see below.

31 Hewett 1985, 155–7, figs 150, 151; Geddes 1999, 319–20. Dendrochronology has confirmed that the timbers are compatible with the historical dating for the construction of the nave (Caple 1999).

32 Hewett 1974, fig. 71.

33 *Ibid.*, 100, fig. 70; Geddes 1999, 330, fig. 4.83.

34 Geddes 1982, 313.

35 The door had five ledges with clasping roves at 20 cm intervals; together they accounted for about forty roves. If there was a bentwood hoop around the perimeter, which seems highly likely, that would have required at least another forty roves to secure it.

36 BL, Cotton Claudius B.IV, fol. 14v. Reproduced in Geddes 1999, fig. 4.1.

37 *Ibid.*, fol. 19. Reproduced in Geddes 1999, fig. 4.2.

38 Rouen, Bibliothèque Municipale, MS 368 (A.27), fol. 2v. For a discussion of the image, see Rodwell 2001, 111–14.

39 BL, Cotton Claudius B.IV, fol. 14r. Reproduced in Geddes 1999, fig. 4.5.

40 Bodleian Library, Oxford, MS Junius 11, p. 66.

41 BL, Stowe 944, fol. 7r.

42 Double-ended straps: *e.g.* Westminster (1); Kempley (3); Stawley (5); Old Woking (6); Merton (8). Illus. in Geddes 1999, figs 4.39, 4.76, 4.17, 4.18.

43 *The Third Life of St Amand*, Bibliothèque Municipale, Valenciennes, MS 500, fol. 51. Illus. in Dodwell 1993, fig. 404.

44 Bodleian Library, Oxford, MS Auct. F.2.13, fol. 4v.

45 Geddes 1999, 52, 373, fig. 4.4.

46 The ironwork was all attached with large-headed nails: twelve in each horizontal strap and ninety-four in the hoop band. The hinges are not shown on Fig. 121 and must have been attached to the back of the door.

47 North doors, *c.* 1208; west front central door, *c.* 1230–60; chapter house crypt, *c.* 1260. Hewett 1985, figs 158, 161, 164.

48 Fernie 1983a, 154–7; 2009; Gem 2009; Woodman 2015.

49 Gem 2009, 169.

50 Geddes 1999, 15; see also Rodwell 2012a.

51 Similar evidence has been found for painting Danish shields red (Underwood 2001, 78–80).

52 Geddes (1999, 15) initially thought the blue-grey paint might be medieval, but later revised this to post-medieval (2006, 54).

53 Brown 1999, 161–7.

54 Rodwell 2010b, 78.

55 Hewett 1985, 169.

56 Bodleian Library: MS Douce 180. Morgan 2007.

57 *Ibid.*, Douce 180, p. 9.

58 Geddes 1999, 360, 372.

59 *Ibid.*, 14, 391–2.

60 *Ibid.*, fig. 3.2.

61 For 'the Piety of Henry III', see Carpenter 2020, chap. 6.

62 Brandon & Brandon 1874, 1, sect. II, pl. 2; Geddes 1999, fig. 4.179.

63 Geddes 1999, 350, fig. 4.73; Orbach *et al.* 2021, 447–8.

64 *Ibid.*, fig. 5.131.

65 *Ibid.*, figs 4.27, 4.28. For Swedish medieval ironwork generally, see Karlsson 1988.

66 There appear to have been several concentric or intersecting circles in the middle two panels (Geddes 1984, 297).

67 Several groups of four nail-holes, in trapezoidal formation, are discernible.

68 A small market was established by 1475, but it failed (Reaney 1935, 338). The settlement is still known as Elmstead Market.

69 RCHME 1916, 47–57. There is no surviving town charter, but the market was in existence by 1216 (Eddy & Petchey 1983, 32, fig. 17.1).

70 Foundations of an earlier apsidal building were discovered beneath the chancel floor, pointing to the existence of a previous church on this site (RCHME 1916, 47).

71  The small door does not have counter-rebated boards, despite being so described by RCHME 1916, 50; Hewett 1974, 99.
72  For discussion of the etymologies, see Reaney 1935, 179–80, 248.
73  Lawson 1993, chap. 1.
74  Swete 1893; Baldwin Brown (1925, 306–7) summarily dismissed Ashdon in favour of Ashingdon, but later reconsidered, as reported by Cobbett (1937, 43). For a full exploration of the evidence, see Rodwell 1993.
75  The alternative of entering Essex via the Colne estuary and following the river valley through Colchester and Castle Hedingham, before joining the Stour valley, is much less likely and can probably be discounted. The Colne has a narrow estuary, with creeks and marshland to either side, and there is no safe or commodious harbour.
76  'We do not know, but we will know.'

# Abbreviations and bibliography

BAA            British Archaeological Association

BAR            British Archaeological Reports

BL             British Library, London

CBA            Council for British Archaeology

LAMAS          London and Middlesex Archaeological Society

MoLAS          Museum of London Archaeological Service

RCHME          Royal Commission on Historical Monuments (England)

Soc. Antiqs    Society of Antiquaries of London

VA             *Vernacular Architecture* (Journal of the Vernacular Architecture Group)

WAM            Westminster Abbey Muniments

Ackermann, R., 1812. *The History of St Peter's, Westminster*. 2 vols. London.

Addyman, P.V., and Goodall, I.H., 1979. 'The Norman Church and Door at Stillingfleet, North Yorkshire', *Archaeologia* 106, 75–105.

Arwidsson, G., and Berg, G., 1982. *The Mästermyr Find: A Viking Age Tool Chest from Gotland*. Nordiska Museet, Stockholm.

Ayres, T., and Tatton-Brown, T., (eds), 2006. *Medieval Art, Architecture and Archaeology at Rochester*. BAA Conference Transactions 28.

Baldwin Brown, G., 1925. *The Arts in Early England, vol. 2: Anglo-Saxon Architecture*. 2nd edn; 1st edn, 1903. Murray, London.

Barnard, E.A.B., (ed.), 1931. *The Prattinton Collections of Worcestershire History*. Evesham.

Baxter, R., 2006. 'The Construction of the West Doorway of Rochester Cathedral', in Ayres and Tatton-Brown, 2006, 85–96.

Benton, G.M., 1936. 'An Early Post-Conquest Doorway and Door found at Elmstead Church', *Transactions of Essex Archaeological Society* (ns) 22, 26–30.

Bettey, J., and Harrison, H., 2004. 'A Twelfth-Century Door in Bristol Cathedral', *Transactions of Bristol & Gloucestershire Archaeological Society* 122, 169–71.

Bettley, J., and Pevsner, N., 2007. *The Buildings of England: Essex*. Yale Univ. Press, New Haven & London.

Bettley, J., Pevsner, N., and Cherry, B., 2019. *The Buildings of England: Hertfordshire*. Yale Univ. Press, New Haven & London.

Brandon, R., and Brandon, J.A., 1874. *An Analysis of Gothick Architecture*. New edn. 2 vols. (1st edn 1847). Batsford, London.

Brandt, L.Ø., Mackie, M., Daragan, M., Collins, M.J., and Gleba, M., 2023. 'Human and Animal Skin Identified by Palaeoproteomics in Scythian Leather Objects from Ukraine', *Plos One*, 18(12), p.e0294129. https://journals.plos.org/plosone/article?id=10.1371/journal.pone.0294129#sec002

Brassington, W.S., 1894. *Historic Worcestershire*. London.

Bridge, M., and Miles, D., 2012. 'Dendrochronologically Dated Doors in Great Britain', *Regional Furniture* 26, 73–103.

Brindle, S., 2010. 'Sir George Gilbert Scott and the Restoration of the Chapter House, 1849–72', in Rodwell and Mortimer 2010, 139–57.

Brooks, A., and Pevsner, N., 2007. *The Buildings of England: Worcestershire*. Yale Univ. Press, New Haven & London.

Brown, S., 1999. *Sumptuous and Richly Adorned: The Decoration of Salisbury Cathedral*. RCHME. HMSO, London.

Bruce-Mitford, R., 1975. *The Sutton Hoo Ship-Burial, vol. 1*. British Museum, London.

Campbell, J.W.P., and Tutton, M., 2020. *Doors: History, Repair and Conservation*. Oxford.

Caple, C., 1999. 'The Durham Cathedral Doors', *Durham Archaeological Journal* 14–15, 131–40.

Carpenter, D., 2020. *Henry III: The Rise to Power and Personal Rule, 1207-1258*. Yale Univ. Press, New Haven & London.

Christensen, A.E., 1982. 'Viking Age Boat-Building Tools', in McGrail 1982, 327–37.

Christy, M., 1925. 'The Battle of "Assandun": Where was it Fought?', *Journal of BAA* (n.s.) 31, 168–90.

Clapham, A.W., 1930. *English Romanesque Architecture, vol. 1: Before the Conquest*. Clarendon, Oxford.

Cobbett, L., 1937. 'Ornament in Hadstock Church, Essex', *Proceedings of Cambridge Antiquarian Society* 37, 43–6.

Coller, D.W., 1861. *The People's History of Essex*. Chelmsford.

Cotte, M., Checroun, E., De Nolf, W., Taniguchi, Y., De Viguerie, L., Burghammer, M., Walter, P., Rivard, C., Salomé, M., Janssens, K., and Susini, J., 2016. 'Lead Soaps in Paintings: Friends or Foes?', *Studies in Conservation* 62(1), 1–22.

[Cromwell, T.K.] Anon., 1818–19. *Excursions in the County of Essex*. 2 vols. London.

Cronyn, J.M., and Horie, C.V., 1985. *St Cuthbert's Coffin: The History, Technology and Conservation*. Dean & Chapter of Durham, Durham.

Darrah, R., 1982. 'Working Unseasoned Oak', in McGrail 1982, 219–23.

Darrah, R., 2009. 'The Tools and Joint Types used in the Construction of the Hemington Bridges', in S. Ripper and L.P. Cooper, *The Hemington Bridges: The Excavation of Three Medieval Bridges at Hemington Quarry near Castle Donington, Leicestershire*. Leicester Archaeol. Monog. 16, 96–113. Leicester.

Dart, J., 1723. *Westmonasterium or The History and Antiquities of the Abbey Church of St Peters Westminster*. London.

Dean, J., and Hill, N., 2014. 'Burn Marks on Buildings: Accidental or Deliberate?', *VA* 45, 1–15.

Dodwell, C.R., 1993. *The Pictorial Arts of the West, 800-1200*. Yale Univ. Press, New Haven & London.

Dunning, G.C., and Goodman, W.L., 1959. 'The Anglo-Saxon Plane from Sarre', *Archaeologia Cantiana* 73, 196–201.

Eddy, M.R., and Petchey, M.R., 1983. *Historic Town in Essex: An Archaeological Survey*. Essex County Council, Chelmsford.

Evans, E., 2000. 'Medieval Churches in Wales', *Church Archaeology* 4, 5–26.

Fern, C., Dickinson, T., and Webster, L., 2019. *The Staffordshire Hoard: An Anglo-Saxon Treasure*. Soc. Antiqs Res. Rep. 80. London.

Fernie, E., 1983a. *The Architecture of the Anglo-Saxons*. Batsford, London.

Fernie, E., 1983b. 'The Responds and the Dating of St Botolph's Hadstock', *Journal of BAA* 136, 62–73.

Fernie, E., 2009. 'Edward the Confessor's Westminster Abbey', in Mortimer 2009, 139–50.

Fiddyment, S., Holsinger, B., Ruzzier, C., Devine, A., Binois, A., Albarella, U., Fischer, R., Nichols, E., Curtis, A., Cheese, E., and Teasdale, M.D., 2015. 'Animal Origin of 13th-century Uterine Vellum Revealed using Non-invasive Peptide Fingerprinting', *Proceedings National Academy of Sciences* 112(49), 15066–71. https://www.pnas.org/doi/10.1073/pnas.1512264112

Fisher, E.A., 1962. *The Greater Anglo-Saxon Churches*. Faber, London.

Geddes, J., 1982. 'The Construction of Medieval Doors', in McGrail 1982, 313–25.

Geddes, J., 1984. 'Decorative Ironwork', in G. Zarnecki, J. Holt and T. Holland (eds), *English Romanesque Art, 1066-1200*, 296–7. Arts Council, London.

Geddes, J., 1999. *Medieval Decorative Ironwork in England*. Soc. Antiqs Res. Rep. 59. London.

Geddes, J., 2006. 'Bishop Gundulf's Door at Rochester Cathedral', in Ayres and Tatton-Brown 2006, 54–60.

Gem, R., 2009. 'Craftsmen and Administrators in the Building of the Confessor's Abbey', in Mortimer 2009, 168–72.

Gibson, A.V.B., and Hewett, C.A., 1983–86. 'St Mary's Church, Little Hormead: Its 12th Century Carpentry', *Hertfordshire Archaeology* 9, 185–9.

Goodburn, D., 1992. 'Woods and Woodland: Carpenters and Carpentry', in Milne 1992, 106–30.

Goodburn, D., 2007. 'Treewrighting and Woodland Management in the 11th and 12th Centuries, in D. Bowsher, T. Dyson, *et al.*, *The London Guildhall: An Archaeological History of a Neighbourhood from Early Medieval to Modern Times*. MoLAS Monog. 36, vol. 2, 302–17.

Goodman, W.L., 1964/1978. *The History of Woodworking Tools*. 2nd edn 1978. Bell & Hyman, London.

Gray, H. St George, 1906–07. 'Notes on Danes' Skins', *Saga-Book of the Viking Club* 5, 218–29.

Hall, R.A., 2014. *Anglo-Scandinavian Occupation at 16–22 Coppergate, York: Defining a Townscape*. The Archaeology of York: Anglo-Scandinavian York 8/5. CBA, York.

Harrison, H., 2007. 'Church Woodwork in England', *Regional Furniture* 21, 53–66.

Harrod, H., 1870. 'On the Crypt of the Chapter House, Westminster Abbey', *Archaeologia* 44, 373–82.

Harvey, J.H., 1987. *English Mediaeval Architects. A Biographical Dictionary down to 1550*. Rev. edn. Sutton, Gloucester.

Hewett, C.A., 1974. *Church Carpentry. A Study based on Essex Examples*. Phillimore, London & Chichester.

Hewett, C.A., 1978. 'Anglo-Saxon Carpentry', *Anglo-Saxon England* 7, 205–29.

Hewett, C.A., 1980. *English Historic Carpentry*. Phillimore, London & Chichester.

Hewett, C.A., 1982. 'Tool-marks on Surviving Works from the Saxon, Norman and Later Medieval Periods', in McGrail 1982, 339–48. Oxford.

Hewett, C.A., 1985. *English Cathedral and Monastic Carpentry*. Phillimore, London & Chichester.

Hewett, C.A., 1988. 'The Jointing of Doors during the Norman Period', *Archaeological Journal* 145, 374–7.

Hewitt, J.B., 1901–02. 'Pembridge Church', *Transactions of Woolhope Naturalists' Field Club 1901-02*, 141–4.

Hill, J., and Woodger, A., 1999. *Excavations at 72-75 Cheapside / 89-93 Queen Street, City of London*. MoLAS, Archaeol. Studies 2. London.

Hope-Taylor, B., 1977. *Yeavering: An Anglo-British Centre of Early Northumbria*. Dept of Environment, Archaeol. Rep. 7. HMSO, London.

Horsman, V., 1988. 'Floors, Doors and Internal Features', in V. Horsman, C. Milne and G. Milne, *Aspects of Saxo-Norman London: I. Building and Street Development*, 85–99. LAMAS, Spec. Paper 11.

Howard, F.E., and Crossley, F.H., 1917. *English Church Woodwork*. Batsford, London.

Hughes, C., 1949. *A Wanderer in North Wales*. Phoenix, London.

James, D., 2012. 'Saw Marks in Vernacular Buildings and their Wider Significance', *VA* 43, 7–18.

Karlsson, L., 1988. *Medieval Ironwork in Sweden*. 2 vols. Stockholm, Sweden.

King, D., 1672. *The Cathedrall and Conventuall Churches of England and Wales, Orthographically Delineated*. 2nd edn. London.

Lawson, M.K., 1993. *Cnut: The Danes in England in the Early Eleventh Century*. Longman, London & New York.

Lethaby, W.R., 1925. *Westminster Abbey Re-examined*. Duckworth, London.

McGrail, S., 1982. *Woodworking Techniques before AD 1500: Papers presented to a Symposium at Greenwich in September 1980*. National Maritime Museum. BAR International Ser. 129. Oxford.

Massey, P., and Reed, P., 2021. 'A Carpenter's Study of the Chapter House Vestibule Door, Westminster Abbey', *Regional Furniture* 35, 1–24.

Melin, K-M., 2018. 'Medieval Counter-Rebated Doors. A Door from the Diocese of Lund compared with the English Examples', *Proceedings of Fifth Conference of Construction History Society*, 249–62.

Miles, D.W.H., and Bridge, M.C., 2005. *The Tree-Ring Dating of the Early Medieval Doors at Westminster Abbey, London*. English Heritage, Centre for Archaeology, Rep. 38/2005. London.

Miles, D., and Bridge, M., 2010a. 'The Chapter House Doors and their Dating', in Rodwell and Mortimer 2010, 251–60.

Miles, D., and Bridge, M., 2010b. 'Tree-Ring Date Lists 2010: List 224', *VA* 41, 102–5.

Miles, D., and Bridge, M.C., 2012. 'Dendrochronologically Dated Doors in Great Britain', *Regional Furniture* 26, 73–101.

Miles, D., and Worthington, M., 2002. 'Tree-Ring Date Lists 2002: List 126', *VA* 33, 81–9.

Miles, D., Worthington, M., and Bridge, M., 2004. 'Tree-Ring Date Lists 2004: List 152', *VA* 35, 95–104.

Miles, D., Worthington, M., and Bridge, M., 2005. 'Tree-Ring Date Lists 2005: List 166', *VA* 36, 88–96.

Miles, D., Worthington, M.J., and Grove, C., 1999. *Tree-ring Analysis of the Nave Roof, West Door and Parish Chest of St Mary, Kempley, Gloucestershire*. Ancient Monuments Laboratory Rep. 36/99. English Heritage, London.

Millett, M., and James, S., 1983. 'Excavations at Cowdery's Down, Basingstoke, Hampshire', *Archaeological Journal* 140, 151–279.

Milne, G., 1992. *Timber Building Techniques in London, c. 900-1400*. LAMAS, Spec. Paper 15. London.

Morant, P., 1768. *The History and Antiquities of the County of Essex*, II. London. Reprinted 1816, Chelmsford.

Morgan, N., 2007. *The Douce Apocalypse. Picturing the End of the World in the Middle Ages*. Bodleian Library, Oxford.

Morris, C.A., 1983. 'A Late Saxon Hoard of Iron and Copper-alloy Artefacts from Nazeing, Essex', *Medieval Archaeology* 27, 27–39.

Mortimer, R. (ed.), 2009. *Edward the Confessor: The Man and the Legend*. Boydell, Woodbridge.

Murray, H., 1983. *Viking and Early Medieval Buildings in Dublin*. BAR 119. Oxford.

Muilman, P., 1770–72. *A New and Complete History of Essex, from a Late Survey, by a Gentleman*. 6 vols. Chelmsford.

Neville, R.C., 1847. *Antiqua Explorata*. Saffron Walden.

Nightingale, J., 1815. *The Beauties of England and Wales*, 10.4. London.

Newcourt, R., 1710. *Repertorium: An Ecclesiastical Parochial History of the Diocese of London*, II. London.

Noppen, J.G., 1936. *Chapter House and Pyx Chamber, Westminster Abbey*. HMSO, London.

Orbach, J., Pevsner, N., and Cherry, B., 2021. *The Buildings of England: Wiltshire*. Yale Univ. Press, New Haven & London.

Ottaway, P., and Cowgill, J., 2009. 'Woodworking, the Tool Hoard and its Lead Containers', in D.H. Evans and C. Loveluck, *Excavations at Flixborough, Vol. 2. Life and Economy at Early Medieval Flixborough, c. AD 600-1000*, 253–77. Oxbow, Oxford & Oakville.

Pepys, S., 1970. *The Diary of Samuel Pepys* (eds R. Latham and W. Matthews), 9 vols. London.

Perkins, J., 1938, 1940, 1952. *Westminster Abbey. Its Worship and Ornaments*. 3 vols. Alcuin Club Coll. OUP, London.

Plant, R., 2006. 'Gundulf's Cathedral', in Ayres and Tatton-Brown 2006, 38–53.

Player, J., 1845. *Sketches of Saffron Walden and its Vicinity*. Saffron Walden.

Quekett, J., 1848. *A Practical Treatise on the Use of the Microscope*. Ballière, London. Rev. edns, 1852 and 1853.

Quekett, J.T., 1849. 'On the Value of the Microscope in the Determination of Minute Structures of a Doubtful Nature, as exemplified in the Identification of Human Skin attached many Centuries ago to the Doors of Churches', *Transactions of Microscopical Society of London* 2, 151–7.

Quekett, W., 1888. *My Sayings and Doings, with Reminiscences of My Life*. London.

Rahtz, P., 1979. *The Saxon and Mediaeval Palaces at Cheddar: Excavations 1960-1962*. BAR 65. Oxford.

RCHME 1916. *An Inventory of the Historical Monuments in Essex. Vol. I, North-West*. HMSO, London.

RCHME 1922. *An Inventory of the Historical Monuments in Essex. Vol. III, North-East.* HMSO, London.

RCHME 1923. *An Inventory of the Historical Monuments in Essex. Vol. IV, South-East.* HMSO, London.

RCHME 1924. *An Inventory of the Historical Monuments in London. Vol. I, Westminster Abbey.* HMSO, London.

RCHME 1934. *An Inventory of the Historical Monuments in Herefordshire. Vol. III, North-West.* HMSO, London.

Reaney, P.H., 1935. *The Place-Names of Essex.* Cambridge Univ. Press, Cambridge.

Reed, R., 1972. *Ancient Skins, Parchments and Leathers.* London & New York.

Reynolds, C., (ed.), 2011. *Surveyors of the Fabric of Westminster Abbey, 1827-1906. Reports and Letters.* Boydell Press, Woodbridge.

Ridgeway, V., (ed.), 2009. *Secrets of the Gardens* [Drapers' Gardens, London]. Pre-Construct Archaeology, Brockley.

Rigold, S.E., 1976. *The Chapter House and Pyx Chamber, Westminster Abbey.* Dept of the Environment. HMSO, London. New edn, 1985, English Heritage, London.

Rodwell, W., 1976. 'The Archaeological Investigation of Hadstock Church, Essex. An Interim Report', *Antiquaries Journal* 56, 55–71.

Rodwell, W., 1993. 'The Battle of *Assandun* and its Memorial Church: A Reappraisal', in J. Cooper (ed.), *The Battle of Maldon: Fiction and Fact*, 127–58. Hambledon Press, London.

Rodwell, W., 2001. *Wells Cathedral: Excavations and Structural Studies, 1978-93.* English Heritage, Archaeol. Rep. 21. London.

Rodwell, W., 2002. *Chapter House and Pyx Chamber, Westminster Abbey.* English Heritage, London.

Rodwell, W., 2009. 'New Glimpses of Edward the Confessor's Abbey at Westminster', in Mortimer 2009, 151–67.

Rodwell, W., 2010a. 'The Chapter House Complex: Morphology and Construction', in Rodwell and Mortimer 2010, 1–31.

Rodwell, W., 2010b. 'Westminster and other Two-Storeyed Chapter Houses and Treasuries', in Rodwell and Mortimer 2010, 66–90.

Rodwell, W., 2012a. 'Appearances can be Deceptive: Building and Decorating Anglo-Saxon Churches', *Journal of BAA* 165, 22–60.

Rodwell, W., 2012b. *The Archaeology of Churches.* 4th edn. Amberley, Stroud.

Rodwell, W., Miles, D., Hamilton, D., and Bridge, M., 2006. 'The Dating of the Pyx Door', *English Heritage Historical. Review* 1, 25–7.

Rodwell, W., and Mortimer R., (eds), 2010. *Westminster Abbey Chapter House: The History, Art and Architecture of 'A Chapter House beyond Compare'*, Soc. Antiqs, London.

Rodwell, W., and Tatton-Brown, T., (eds), 2015. *Westminster: The Art, Architecture and Archaeology of the Royal Abbey and Palace.* BAA Conference Transactions 39(1).

Rogerson, A., and Dallas, C., 1984. *Excavations in Thetford, 1948-59 and 1973-80*, East Anglian Archaeol. Reps 22. Norwich.

Scott, G.G., 1860. 'Gleanings from Westminster Abbey', *Gentleman's Magazine.* New ser. 8, 128–37, 250–7, 351–61, 462–9, 577–84; new ser. 9, 33–40 (in six parts).

Scott, G.G., 1861. *Gleanings from Westminster Abbey.* 2nd edn 1863. Parker, Oxford & London.

Sheppard, T., 1939. 'Viking and other Relics at Crayke, Yorkshire', *Yorkshire Archaeological Journal* 34, 273–81.

Stanley, A.P., 1868/1882. *Historical Memorials of Westminster Abbey.* 1st & 5th edns. London.

Stukeley, W., 1724. *Itinerarium Curiosum* I. 2nd edn, 1776. London.

Swanton, M.J., 1976 '"Dane skins": Excoriation in Early England', *Folklore* 87, 21–8.

Swete, H.B., 1893. 'On the Identification of Assanduna with Ashdon', *Transactions of Essex Archaeological Society* ns 4, 5–10.

Taylor, H.M., and Taylor, J., 1965 & 1978. *Anglo-Saxon Architecture.* 3 vols. Cambridge Univ. Press, Cambridge.

Theophilus: trans. & ed., Dodwell, C.R., 1961. *Theophilus, De Diversis Artibus*. London & Edinburgh.
Theophilus: trans. & ed., Hawthorne, J.G., and Smith, C.S., 1979. *On Divers Arts*. New York.
Thomas, G., 2010. *The Later Anglo-Saxon Settlement at Bishopstone: A Downland Manor in the Making.* CBA Res. Rep. 163. York.
Thomas, G., 2013. 'Life before the Minster: The Social Dynamics of Monastic Foundation at Anglo-Saxon Lyminge, Kent', *Antiquaries Journal* 93, 109–45.
Thomson, E., 1955. *Clifton Lodge*. Hutchinson, London.
Tristram, E.W., 1944. *English Medieval Wall Painting: The Twelfth Century*. The Pilgrim Trust. OUP, Oxford.
Tweddle, D., Biddle, M., and Kjølbye-Biddle, B., 1995. *Corpus of Anglo-Saxon Stone Sculpture, Vol. IV: South-East England*. British Academy. OUP, Oxford.
Tyack, G.S., 1898. 'Human Skin on Church Doors', in W. Andrews (ed.), *The Church Treasury of History, Custom, Folk-Lore, etc.*, 158–67. London.
Tyers, I., 1996. *Tree-ring Analysis of Timbers from the Stave Church at Greensted, Essex*. Ancient Monuments Laboratory, Rep. 14/96.
Underwood, R., 2001. *Anglo-Saxon Weapons and Warfare*. Tempus, Stroud.
Way, A., 1847. *Catalogue of Antiquities and Miscellaneous Curiosities in Possession of the Society of Antiquaries*. Soc. Antiqs, London.
Way, A., 1848. 'Some Notes on the Tradition of Flaying, inflicted in Punishment of Sacrilege', *Archaeological Journal* 5, 185–92.
Williams, R.J., 2011. 'Other Saxon and Medieval Finds' and 'Wooden Objects from Northfleet', in P. Andrews, L. Mepham, J. Schuster and C.J. Stevens, *Settling the Ebbsfleet Valley: High Speed 1 Excavations at Springhead and Northfleet, Kent*, 4. Oxford Wessex Archaeology, Oxford.
Wilson, D.M., 1976. 'Craft and Industry', in D.M. Wilson (ed.), *The Archaeology of Anglo-Saxon England*, 253–81. Methuen, London.
Woodman, F., 2015. 'Edward the Confessor's Church at Westminster: An Alternative View', in Rodwell and Tatton-Brown 2015, 61–8.
Wright, T., 1836. *History and Topography of the County of Essex*. 2 vols. London.

# Index

Page numbers in italics refer to illustrations